Lecture Notes in Computer Science　　7715

Commenced Publication in 1973
Founding and Former Series Editors:
Gerhard Goos, Juris Hartmanis, and Jan van Leeuwen

Editorial Board

T0171822

Shlomi Dolev   Mihai Oltean (Eds.)

Optical Supercomputing

4th International Workshop, OSC 2012
in Memory of H. John Caulfield
Bertinoro, Italy, July 19-21, 2012
Revised Selected Papers

 Springer

Volume Editors

Shlomi Dolev
Ben-Gurion University of the Negev
Department of Computer Science
84105 Beer Sheva, Israel
E-mail: dolev@cs.bgu.ac.il

Mihai Oltean
Babeş-Bolyai University
Faculty of Mathematics and Computer Science
Department of Computer Science
Kogălniceanu 1
Cluj-Napoca, 40084, Romania
E-mail: moltean@cs.ubbcluj.ro

ISSN 0302-9743 e-ISSN 1611-3349
ISBN 978-3-642-38249-9 e-ISBN 978-3-642-38250-5
DOI 10.1007/978-3-642-38250-5
Springer Heidelberg Dordrecht London New York

Library of Congress Control Number: 2013940583

CR Subject Classification (1998): F.1, B.4.3, C.5.1

LNCS Sublibrary: SL 1 – Theoretical Computer Science and General Issues

Typesetting: Camera-ready by author, data conversion by Scientific Publishing Services, Chennai, India

Printed on acid-free paper

Springer is part of Springer Science+Business Media (www.springer.com)

Preface

OCS, the International Workshop on Optical SuperComputing, is a forum for research presentations on all facets of optical computing for solving hard computation tasks. Optical computing devices have the potential to be the very next computing infrastructure. The frequency limitations and cross-talk phenomena, as well as soft-errors of electronic devices on one hand, and the natural parallelism of optical computing devices, as well as the advances in fiber optics and optical switches, on the other hand, make optical computing commercialable. The focus of OCS, on research in the theory, design, specification, analysis, implementation, or application of optical supercomputers. Topics of interest include, but are not limited to: design of optical computing devices, electoroptics devices for interacting with optical computing devices, practical implementations, analysis of existing devices and case studies, optical and laser switching technologies, applications and algorithms for optical devices, alpha practical, x-Rays, and nano-technologies for optical computing. The first edition of OSC was held on August 26, 2008, in Vienna, Austria, co-located with the 7th International Conference on Unconventional Computing; the second OSC was held during November 18–20, 2009, in Bertinoro, Italy; the third OSC was held during November 17–19, 2010, in Bertinoro, Italy, and the fourth during July 19–20, 2012, in Bertinoro, Italy.

This volume contains 13 contributions selected by the Program Committee and six invited papers. All submitted papers were read and evaluated by the Program Committee members, assisted by external reviewers. We are grateful to the EasyChair system for assisting the reviewing process.

OSC 2012 was organized in cooperation with OSA, the Optical Society of America. The support of Ben-Gurion University and Babeş-Bolyai University is also gratefully acknowledged.

January 2013

Shlomi Dolev
Mihai Oltean

Organization

OSC, the International Workshop on Optical SuperComputing, is a forum for research presentations on all facets of optical computing. OSC 2012 was organized in cooperation with the OSA, the Optical Society of America, and in conjunction with the International Conference on Unconventional Computation.

Founding Steering Committee

H. John Caulfield	Fisk University, USA
Shlomi Dolev	Ben-Gurion University of the Negev, Israel
Mihai Oltean	Babeş-Bolyai University, Romania

Organizing Committee

Program Chairs	Shlomi Dolev, Ben-Gurion University of the Negev, Israel
	Mihai Oltean, Babeş-Bolyai University, Romania

Program Committee

George Barbastathis	MIT, USA
Antonella Bogoni	CNIT, Italy
Yeshaiahu Fainman	University of California, San Diego, USA
Debabrata Goswami	Indian Institute of Technology Kanpur, India
Tobias Haist	Universität Stuttgart, Germany
Zhanghua Han	University of Alberta, Canada
Boris Kryzhanovsky	Center of Optical Neural Technologies, SRISA RAS, Moscow
Alstair Mcaulay	Lehigh University, Bethlehem, Pennsylvania
Kouichi Nitta	Kobe University, Japan
Mihai Oltean	Babeş-Bolyai University, Romania
Wolfgang Osten	Universität Stuttgart, Germany
Haldun Ozaktas	Bilkent University, Ankara, Turkey
Joseph Rosen	Ben-Gurion University, Israel
Sukhdev Roy	Dayalbagh Educational Institute, India
Joseph Shamir	Technion Institute, Israel
Dan Tamir	Texas State University, USA
Damien Woods	Caltech, USA
Zeev Zalevsky	Bar-Ilan University, Israel
Xinliang Zhang	Huazhong University of Science and Technology, China

Table of Contents

Obituaries

Quantum Optical Transient Encryption and Processing

Zeev Zalevsky[1], David Sylman[1], and H. John Caulfield[2]

[1] Faculty of Engineering, Bar-Ilan University, 52900 Ramat-Gan, Israel
zalevsz@biu.ac.il
[2] Alabama A&M University Research Institute, Normal Al 35782, Alabama, USA

Abstract. This paper is written in the memory of H. John Caulfield, a unique breaking through scientist that was known for his capability of thinking "out of the box" and which has made large number of significant scientific contributions in many optics related fields. In the last few years of his carrier, John has invested some time in the field of optical quantum computing and encryption. In this paper we wish to describe one of such directions investigated by John and dealing with optical quantum encryption and processing.

Keywords: Interferometry, encryption, polarization based encoding.

1   Introduction

This paper is written in the memory of H. John Caulfield, a unique and fruitful scientist, very special colleague and a warm friend. John had a scientifically breaking through carrier during which he has published more than 300 papers and book chapters. His large scientific diversity and huge multidisciplinary background allowed him to make a significant contribution in large variety of optical fields. Some of his papers became bench marks others revolutionized some of the modern fields of optics. His publications contributed mainly in fields as: holography recording and storage (including materials), metrology and optical testing, optical data processing and computing, neural networks and optical fuzzy logic schemes, optical transformations such as Wavelets and fractional Fourier transform. Some of his scientific throughputs made exceptional impact e.g. the papers in Refs. [1,2]. In this paper we wish to describe a work of John made in the field of optical quantum computing and encryption [3].

For over 1000 years, people have used encoding and decoding to send secret messages. What has changed is the way that the methods have improved from the Chinese remainder method to entanglement-based quantum computing. Changing that universal method for the first time in the last millennium produces signals that turn irretrievably into mixtures of signal and noise upon detection. Decoding does not work, as no encoding is used. To receive the pure signal, the user must do the proper pre-detection operations at the right times in an optical system online with the transmitted beam. Here we show the basic concept along with a laboratory demonstration using the simplest possible means - polarized light in free space.

S. Dolev and M. Oltean (Eds.): OSC 2012, LNCS 7715, pp. 1–6, 2013.

The pre-entanglement quantum mechanics of Schrödinger can be used to send information in such a way that if it is detected, the message is irreversibly mixed with noise – a use for the irreversible loss of information upon detection. An intercepted signal can never be decoded, because it was never encoded. To render the signal readable, the receiver must be in line with the beam and perform the proper pre-detection operations on the light and do so at the right time. The importance of this form of secure communication can be seen in several ways: (a). The sophisticated computer analysis the NSA and others do will be completely ineffective with the irreversible loss of the needed information. Decoding does not work absent encoded messages to decode. Decoding what is not encoded is impossible. (b). This is the only current alternative to encoding/decoding for secure communication. When there is only one approach, inventing an alternative usually proves very difficult, because everyone has the same implicit assumptions. When there are two totally unrelated approaches, the community tends to invent more. As now, the assumptions are explicit and it can become a possible thing to change the current paradigm into yet another approach beyond the first two ways to do secure communications.

Optical encryption techniques have several advantages over conventional digital encryption due to its high space bandwidth product and high speed data processing capability. The option of using the degrees of freedom of light such as amplitude, phase and polarization yield several different optical encryption methods that have been suggested before [4]. Refregier and Javidi [5] had proposed a phase method based on using double random phases for encryption. Other optical encryption methods were proposed after this paper and used an addition of a third dimension to the random phase mask [6], full random phase encryption [7], using the fractional Fourier domain [8,9], spectral [10] and polarization keys [11-13] for encryption as well as asymmetric form for optical security encryption [14].

2   Technical Description

An appropriately descriptive name for the new way to send and receive signals as it is presented in this paper is QUantum Optical Transient Encryption – QUOTE. The "Quantum Optical" component of this technique is obvious. This is a classical quantum measurement. The only encryption possible must be done live. Recording the signal does not produce an encrypted signal. If the user does not apply the right pre-detection operation, the detected signal will be worthless to anyone. That is what is meant by "Transient Encryption". Nothing that can be recorded and carried back to experts and their supercomputers will be of value to anyone, except that trying to decode messages can waste a lot of time. That may be something the transmitter/receiver would want to accomplish as a side effect of their transmission and reception of a message.

The essential concepts are quite simple. Figure 1 shows all there is to QUOTE. This is the basically a demonstration of an approach of sending secret information that does not use encryption at all.

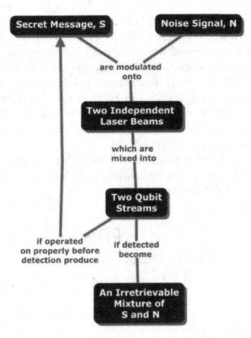

Fig. 1. The basic concept of QUOTE is shown here. The signal S and the noise N are mixed by beam splitters. The two resulting beams contain a mixture of S and N that, if detected, turn into an irreversible mixing of S from N. Operating on the incoming signals with another interferometer at the receiver end produces S at one output and N at the other.

Instead, it mixes the amplitudes of two optical beams coherently, so that the two beams, call them S and N for are in phase quardrature. Detection of either or both beams results a reading of $|S|^2 + |N|^2$. This has the very useful property of being impossible to separate back into the signal S and a noise N if it is detected. The needed phase information is destroyed by the detection – Schrödinger's famous "irreversible loss of information on detection". So the directly detected beam or beams are worthless to anyone. They cannot separate S from N, because the needed information is lost and cannot be recovered. This is the problem Dennis Gabor solved for wave fronts bearing spatial patterns. Thus, this is equivalent to a type of temporal holography. But it should not be confused with other temporal holography methods such as photon echo methods.

Note that the problem posed by that loss of information applies not only to interlopers but also to the intended recipient. No amount of post detection analysis can restore the signal S. That is precisely what Schrödinger meant in his frequent referring to the irreversibility of loss of information. There is no code and thus there can be no successful decoding. The vast efforts in encryption and decryption become irrelevant in this case. Not even quantum computers can decode what was never encoded.

But that recipient can do the proper pre-detection operation that separates S and N and thus allows the reading of $|S|^2$ and the discarding of $|N|^2$. Polarization is the easiest way to demonstrate secret messaging with no codes. It is almost certainly not the best way, but it does suffice to illustrate physically the basic concept just described.

A Mach-Zehnder Interferometer (MZI) is used to add external high intensity noise making the information signal undetectable. By proper combination of the two arms of the interferometer the noise and the signal can be re-separated at the output port. The proposed interferometer uses polarized beam splitter (PBS) thus in our case only if there is a polarizer in the correct direction the noise will be filtered out from the signal. However, similar configuration can be realized with regular beam splitters (BS) and then instead of using the polarization domain it is the phase related correlation obtained due to the coherence of the light that is used for the encryption and later on for the information decryption.

In the experimental validation we confirmed the ability to transmit signal and noise through the system (see Fig. 2) that was adjusted and later on to separate them back again with the proper pre-detection operation.

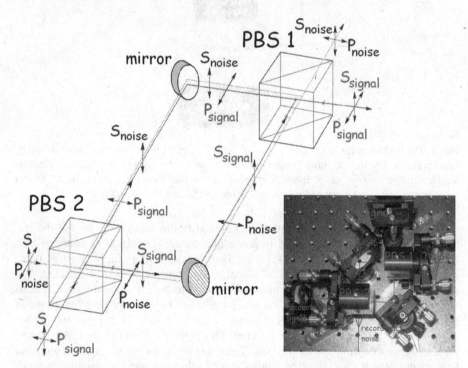

Fig. 2. The schematic sketch and the image of the constructed experimental setup (lower right edge)

We examined the system with polychromatic light (white light) illumination. The results are seen in Fig. 3. In Fig. 3(a) we present the reconstructed signal and in Fig. 3(b) the reconstructed noise. In Fig. 3(c) we show the mixture of the noise and the signal obtained in one of the intermediate outputs of the PBS. The ability of constructing the signal was also obtained after significantly reducing the intensity of the signal by adjusting a neutral density (ND) filter at the signal entry. One may see that less than 1% of the noise can be seen in Fig. 3(d). There signal's transmittance was significantly lower than the noise level and it was equal to 0.09, 0.03 and 0.009 (9%, 3% and 0.9% from the noise intensity), respectively.

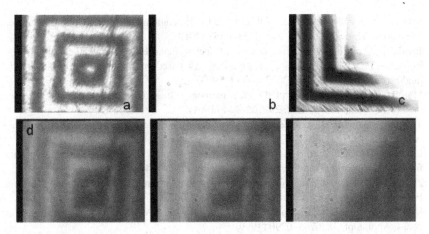

Fig. 3. The experimental results for white light illumination. (a). The reconstructed signal. (b). The reconstructed noise distribution. (c). The mixture of the noise and the signal. (d). Reconstruction of the signal which is significantly lower than the noise level. Signal's transmittance is (from left to right): 0.09, 0.03, 0.009.

In conclusion, the general concept of mixing and unmixing of signal and noise in the optical domain in such a way that simple measurement of either or both of the mixed signals results in readings that have lost the information needed for unmixing has been experimentally demonstrated with the expected good results. Thus, this approach joins the old and familiar method of encryption and decryption as a viable way to send secret information while preventing the message from being read by unintended recipients.

3 Conclusions

In this paper we have demonstrated a simple method for optical transmission of secret messages using the general principle of mixing and demixing of amplitudes of a signal S and a noise pattern N. This was based upon Mach-Zehnder interferometer. The experimentally demonstrated mixing and unmixing of signal and deliberate noise was demonstrated with polarization coding where the signal and the noise were mixed together and later on separated, by the interferometer, based upon their polarization state.

This state can randomly vary in time which will make the decryption very complicated for unauthorized source especially when the noise level is significantly stronger than the information signal. The use of images as signals (instead of time varying signals) was a choice made to show the results easily.

References

1. John Caulfield, H.: Transmission through a tapered quartz tube in the laser near field. Nature 208, 773 (1965)
2. Greguss, P., Caulfield, H.J.: Multiplexing Ultrasonic Wave Fronts by Holography. Science 177, 422 (1972)

3. Sylman, D., Zalevsky, Z., Caulfield, H.J.: Entanglement based Optical Transient Encryption. Optics Commun. 283, 4551–4557 (2010)
4. Javidi, B.: Securing information with optical technologies. Phys. Today 50, 27–32 (1997)
5. Refregier, P., Javidi, B.: Optical image encryption based on input plane and Fourier plane random encoding. Opt. Lett. 20, 767–769 (1995)
6. Matoba, O., Javidi, B.: Encrypted optical memory system using three-dimensional keys in the Fresnel domain. Opt. Lett. 24, 762–764 (1999)
7. Towghi, N., Javidi, B., Luo, Z.: Fully phase encrypted image processor. J. Opt. Soc. Am. A 16, 1915–1927 (1999)
8. Liu, S.T., Mi, Q.L., Zhu, B.H.: Optical image encryption with multistage and multichannel fractional Fourier-domain filtering. Opt. Lett. 26, 1242–1244 (2001)
9. Zhang, Y., Zheng, C.H., Tanno, N.: Optical encryption based on iterative fractional Fourier transform. Opt. Commun. 202, 277–285 (2002)
10. Matoba, O., Javidi, B.: Encrypted optical storage with wavelength-key and random phase codes. Appl. Opt. 38, 6785–6790 (1999)
11. Javidi, B., Nomura, T.: Polarization encoding for optical security systems. Opt. Eng. 9, 2439–2443 (2000)
12. Tan, X., Matoba, O., Okada-Shudo, Y., Ide, M., Shimura, T., Kuroda, K.: Secure optical memory system with polarization encryption. Appl. Opt. 40, 2310–2315 (2001)
13. Matoba, O., Javidi, B.: Secure holographic memory by double random polarization encryption. Appl. Opt. 43, 2915–2919 (2004)
14. Qin, W., Peng, X.: Asymmetric cryptosystem based on phase-truncated Fourier transforms. Opt. Lett. 35, 118–120 (2010)

Parallel Processing for Prime Factorization with Spatial Amplitude Modulation in Optics

Kouichi Nitta, Takashi Kamigiku, Takeshi Nakajima, and Osamu Matoba

Department of Systems Science
Graduate of System Informatics, Kobe University
Rokkodai-cho 1-1, Nada-ku, Kobe, Hyogo, 657-8501, Japan
{nitta,matoba}@kobe-u.ac.jp

Abstract. An optical method for prime factorization is modified. The procedure in the original method is similar to that with quantum computing. In this repot, some differences between the quantum solution and the method are discussed. And, improvement for our method is proposed based on the discussion.

Keywords: Prime factorization, Spatial parallel processing, Amplitude modulation.

1 Introduction

Effective methods for computationally hard problems are required in computer science. Ultra-scale parallel processing with quantum or molecular computing is considered to be attractive for such problems.

Optical information processing is also may have potential power to contribute to such research topics [1] and some solutions have been reported. In Refs. [2,3] methods for the travelling salesman problem. One method is use of optical correlation and the other is based on white light interferometry.

Prime factorization also requires huge computational costs [4]. We have developed an optical method for prime factorization [5~7]. In the method, phase modulation of light-wave is utilized for parallel processing in modulo operations. Moreover, we have proposed another optical procedure [8]. In the procedure, amplitude modulation is the main operation for prime factorization. This operation gives us distribution similar to modulo exponentiation. Mathematical processing in our procedure is almost same as that of Shor's quantum algorithm except for some differences.

In this report, these differences are indicated and analyzed. In accordance with the analysis, two ideas to improve our procedure are presented and we discuss on effectiveness of the ideas.

Section 2 shows a procedure of an algorithm for prime factorization [9] and the original method briefly. Section 3 describes an extended method for phase modulation [10]. In section 4, we propose a modified method based on thresholding. In section 5, usefulness of the modified method is verified by numerical simulation.

S. Dolev and M. Oltean (Eds.): OSC 2012, LNCS 7715, pp. 7–14, 2013.

2   Prime Factorization with Optical Amplitude Modulation

Let us consider that prime factors p and q. Here the both two numbers are assumed to be unknown and different each other. N is also given by product of p and q. The following algorithm is to derive p and q from N.

In the algorithm, modulo exponentiation defined by Eq. (1) is calculated in $0 \leq x \leq 2N^2\text{-}1$.

$$f(x) = a^x \bmod N \tag{1}$$

Here we should select a natural number a which is satisfied with the conditions as defined in Eqs. (2) and (3).

$$1 < a < N \tag{2}$$

$$\gcd(a, N) = 1 \tag{3}$$

With the Fourier analysis, the period of $f(x)$ can be obtained. If the period defined as r is an even number, p and q are derived by the following equations.

$$p = \left(a^{r/2} - 1, N\right), \ q = \left(a^{r/2} + 1, N\right) \tag{4}$$

In Shor's algorithm, specific parallel processing is utilized to obtain $f(x)$. This processing is based on Shonhage Strassen method. At first a set of $l(i)$ is prepared for $i = 0, 1, 2, \ldots n\text{-}1$

$$l_i = a^{2^i} \bmod N \tag{5}$$

With the set, parallel operations to obtain $f(x)$ is in accordance with Eq. (6).

$$f(x) = l_0^{x_0} l_1^{x_1} l_2^{x_2} \ldots l_i^{x_i} \ldots l_{n-1}^{x_{n-1}} \bmod N \tag{6}$$

x_i shows an i'th bit value of x. For example, a set of $(x_0, x_1, x_2, x_3, x_4)$ is $(0,1, 0, 1, 0)$ at $x=10$ and $n=5$. Fig. 1 shows a relations between $f(x)$ and l_i's.

In Ref. [8] an optical procedure to execute operations similar to Eq. (6). Fig. 2 shows a schematic diagram of an optical system for the procedure. This system consists of a sequence of imaging optics. Spatial light modulators for amplitude modulation are put on image planes. Transmittance distribution at the pixel x on the SLM_i is represented as following equation.

$$t_i(x) = \frac{l_i^{x_i}}{\max\{l_0, l_1, \ldots l\}} \tag{7}$$

As results of optical parallel processing, patterns corresponding with $f(x)$ described in Eq. (7) is detected on the output plane.

$$g(x) = \prod_{i=0}^{n-1} l_i^{x_i} \tag{8}$$

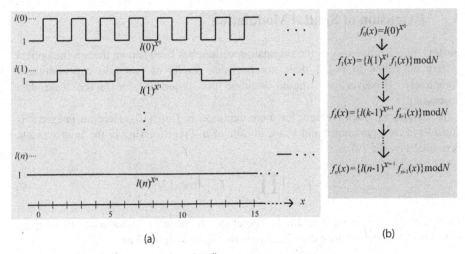

Fig. 1. (a) Structure of $l(i)^{x_i}$ and (b) the procedure to derive $f(x)$

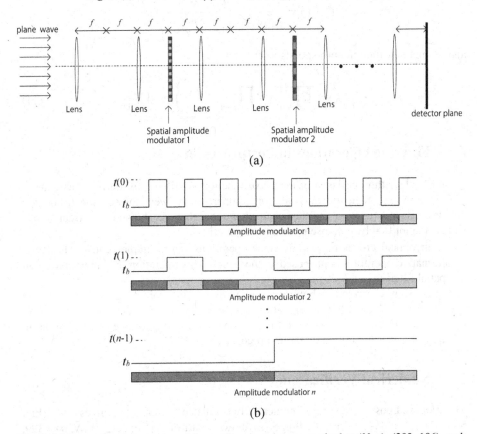

Fig. 2. (a) Parts of power spectrum of a profile given by our method at $(N, a)=(203, 106)$, and (b) that of $f(x)$

3   Extension of Spatial Modulation

In Ref. [8], effectiveness of the original procedure has been shown though the optical system described in Fig 1 does not provides results of modulo exponentiation completely. However, we should improve our procedure for larger scale data processing.

A method for improvement has been reported in Ref. [10]. Here an integer k is considered as a parameter and k is a divisor of n. Preprocessing in the improvement is replaced by Eq. (9).

$$l'_j = \left(\prod_{i=jk}^{(i+1)k-1} l_j^{x_j} \right) \bmod N \qquad (9)$$

From Eq. (9), j is from 0 to k/n-1. Therefore, the number of sequence of imaging optics is k/n. Transmittance distribution on the SLM$_j$ is defined as

$$t'_j(x) = \frac{l_j}{\max(l'_0, l'_1, ..., l'_{n/k-1})} \qquad (10)$$

and, output distribution corresponds with

$$g'_j(x) = \prod_{i=0}^{n/k-1} \left\{ \left(\prod_{i=jk}^{(i+1)k-1} l_j^{x_j} \right) \bmod N \right\} \qquad (11)$$

4   Filtering Operations in Postprocessing

In Shor's algorithm, collapse of the wave function is utilized for derive r accurately. Fig. 3 shows a schematic diagram for comparison between power spectrum with collapse of wave function and the spectrum without that. As shown in the figure, peak signal is amplified by collapse in quantum operations.

In our optical processing, quite same operations can be implemented. Therefore, an alternative operation is proposed in the paper. This operation is simple and based on spatial filtering.

Fig. 4 shows a relation between spatial distribution obtained with the optical system and that after filtering operation. In the operation, thresholding is applied to pixel values. From Figs. 4 and 5, a filtering operation seems to give effects similar to collapse of wave function in quantum process.

5   Numerical Verification

The proposed post processing is numerically estimated to verify usefulness of it. Fig. 5 shows an example of results. Fig. 5 (a) shows a part of $g(x)$ in case of (N, a)=(899, 417). Fig. 5 (b) depicts the results of filtering operations to $g(x)$. Figs. 5 (c) and (d) show auto correlation of profile described in Figs 5 (a) and (b), respectively.

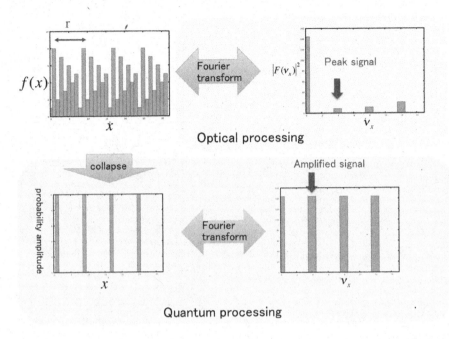

Optical processing

Quantum processing

Fig. 3. Diagram to show effectiveness of collapse of the wave function in prime factorization

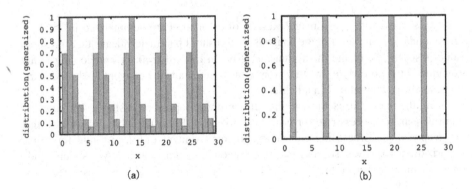

(a)

(b)

Fig. 4. (a) An example of distribution of f(x), and (b) result of filtering with threshold value is 0.7~0.9

In this case, r is 420. From the graphs, the peak signal in Fig. 5(d) is stronger than that in Fig. 5(c). From the comparison, it is found that the proposed postprocessing is effective for prime factorization. Note that both procedures with and without postprocessing give the correct period in this case.

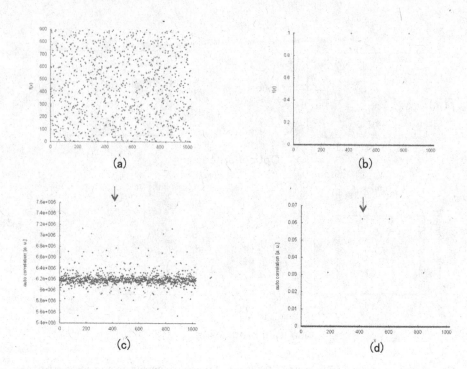

Fig. 5. Results of numerical analysis at (N, a)=(899, 417). (a) profile of $g(x)$, (b) that after thresholding, (c) auto-correlation of $g(x)$, and (d) auto-correlation of (b).

Next, success rate is estimated. Results of estimation are summarized in Table 1. Three kinds of N are used in the analysis. A threshold parameter means that threshold value is set to $\alpha\%$ of the maximum intensity. In case of α=0, postprocessing is not executed in the procedure. In other words, the case means the conventional procedure. In the analysis, k is set to 2 in all cases.

From the table, it is shown that the proposed postprocessing contributes to improvement of success rate for prime factorization. From the table, it is shown that the proposed postprocessing contributes to improvement of success rate for prime factorization. However, success rate is not enough to practical use. We should study on a scheme to optimize the values of k and α, respectively. That is considered to be one of the future issues.

Table 1. Results of estimation to investigate success rate. A and B show the number of success and that of trail, respectively

Bit length of N	Threshold parameter	A/B
10 ($N = 899$)	0%	18/179
	12.5%	21/179
	25%	24/179
	50%	25/179
20 ($N = 944423$)	0%	5/10000
	12.50%	8/10000
	25%	7/10000
	50%	16/10000
30 ($N = 802021439$)	0%	0/10000
	12.50%	1/10000
	25%	1/10000
	50%	2/10000

6 Summary

Post-processing in an optical procedure for prime factorization has been proposed. This post-processing is based on thresholding to intensity distribution obtained with the optical processing. The purpose of the proposed processing is implementation of operations corresponding with collapse of wave function in Shor's quantum algorithm. Effectiveness of it has been clarified by numerical analysis.

We comment on optical implementation for our system. Fig. 2 (a) shows schematic diagram for explaination. However, the system shown in the figure is not so practical in terms of alignment. An optical feedback system seems to be effective to implement a sequence of modulation in our method.

We should construct an experimental system with the optical procedure. Also, improvement for larger scale prime factorization based on the proposed procedure is required.

References

1. Oltean, M.: Solving the Hamiltonian path problem with a light-based computer. Natural Computing 7, 57–70 (2008)
2. Haist, T., Osten, W.: An Optical Solution for the Traveling Salesman Problem. Opt. Express 15, 10473–10482 (2007)

3. Shaked, N.T., Messika, S., Dolev, S., Rosen, J.: Optical solution for bounded NP-complete problems. Appl. Opt. 46, 711–724 (2007)
4. Shamir, A.: Factoring Large Numbers with the TWINKLE Device (Extended Abstract). In: Koç, Ç.K., Paar, C. (eds.) CHES 1999. LNCS, vol. 1717, pp. 2–12. Springer, Heidelberg (1999)
5. Nitta, K., Matoba, O., Yoshimura, T.: Parallel processing for multiplication modulo by means of phase modulation. Appl. Opt. 47, 611–616 (2008)
6. Nitta, K., Katsuta, N., Matoba, O.: An Optical Parallel System for Prime Factorization. Jpn. J. Appl. Phy. 48, 09LA02-1–09LA02-5 (2009)
7. Nitta, K., Katsuta, N., Matoba, O.: Improvement of a system for prime factorization based on optical interferometer. In: Dolev, S., Oltean, M. (eds.) OSC 2009. LNCS, vol. 5882, pp. 124–129. Springer, Heidelberg (2009)
8. Nitta, K., Matoba, O.: An optical system for prime factorization based on parallel processing. In: Dolev, S., Oltean, M. (eds.) OSC 2010. LNCS, vol. 6748, pp. 10–15. Springer, Heidelberg (2011)
9. Shor, P.: Algorithms for quantum computation: Discrete logarithms and factoring. In: Proc. 35th Ann. Symp. on Foundations of Comput. Sci., vol. 1898, pp. 124–134 (1994)
10. Kamigiku, T., Nitta, K., Matoba, O.: Improvement of modulation patterns in a method for parallel modulo exponentiation with optical amplitude modulation. Technical Digest of Moc 2011, H-80 (2011)

An Optical Polynomial Time Solution for the Satisfiability Problem

Sama Goliaei and Saeed Jalili

Electrical and Computer Engineering Department, Tarbiat Modares University,
Tehran, Iran
{goliaei,sjalili}@modares.ac.ir

Abstract. In this paper, we have used optics to solve the satisfiability problem. The satisfiability problem is a well-known NP-complete problem in computer science, having many real world applications, which no polynomial resources solution is found for it, yet. The provided method in this paper, is based on forming patterns on photographic films iteratively to solve a given satisfiability problem in efficient time. The provided method requires polynomial time, but, exponential length films and exponential amount of energy to solve the satisfiability problem.

Keywords: Optical Computing, Unconventional Computing, Optical Problem Solving, Satisfiability problem.

1 Introduction

The problem of finding if a given boolean formula is satisfiable or not is referred as the satisfiability (SAT) problem [1]. A boolean formula in propositional logic is an expression written using only variables, parenthesis, and logical operations "and", "or", "not". A boolean formula is satisfiable if and only if it is possible to assign a value ("true" or "false") to each variable in such a way that the whole formula is evaluated to "true".

The satisfiability problem is an important problem in combinatorics as it is the first known NP-complete problem and every other NP-complete problem can be reduced to a satisfiability instance [2]. The satisfiability problem has also wide range applications in model checking to verify temporal logic properties of systems [3] and appears in many software and hardware verification systems [4]. A huge number of solutions for the satisfiability problem is provided, but no polynomial resources solution taking polynomial amount of resource such as time, space, and energy is found yet [5].

Beside electronic computers, unconventional computing has grown from many years ago in order to bring more computational capabilities in computation, by using various natural phenomena rather than just electronics. Many real-world hard problems are already investigated in unconventional computing and some good results are obtained. Different versions of the satisfiability problem are also investigated in different branches of unconventional computing and new solutions

S. Dolev and M. Oltean (Eds.): OSC 2012, LNCS 7715, pp. 15–24, 2013.

based on different natural phenomena are already provided, such as quantum computing [6], DNA computing [7], and membrane computing solutions [8].

Optical computing as a branch of unconventional computing performs computation based on physical properties of light. Special properties of light such as ability of splitting a light ray into several light rays, high parallel nature, and existence of different wavelengths in a simple light ray are recently used to provide new solutions for several NP-complete problems, specially the satisfiability problem. In some works different wavelengths in a light ray are used to solve a restricted version of the satisfiability problem, 3-SAT problem [9,10], where different wavelengths are considered as possible solutions for the 3-SAT problem and prisms and beam splitters are used to select the proper solutions. In some other works, the satisfiability problem is solved by making delays on light ray motion [11,12] where light rays reaching a destination in a specific time represent a proper solution for the satisfiability problem. Another works solve the satisfiability problem by reducing the light intensity [11], and xeroxing on plastic sheets [13]. Although these solutions have many restrictions and difficulties in practice, they bring the idea that optical approaches may yield to obtain more efficient solutions to the satisfiability problem.

In this paper, we have provided a novel optical solution for the satisfiability problem introducing a new design for optical devices and taking polynomial time to solve each problem instance. We have previously provided optical designs based on using optical sensitive sheets to solve the dominating set problem [14] and the graph 3-colorability problem [15]. In the next section, the satisfiability problem is defined precisely. In section 3, the novel optical solution is provided. The complexity of the provided solution is explained in section 4. In section 5 advantages and disadvantages of the provided solution is investigated. The conclusion and future works are finally provided in section 6.

2   The Satisfiability Problem

The satisfiability (SAT) problem in propositional logic is referred to a decision problem asking if a given formula over n variables is satisfiable or not. In the other words, in each instance of the satisfiability problem, a formula F composed of n variables $x_1, \cdots, x_n$, parenthesis, and three logical operations "and", "or", and "not" is given, and the question is weather it is possible to assign either true ("1") or false ("0") to each variable in such a way that F be evaluated to true ("1").

In this paper, we assume that the satisfiability formula is given in conjunctive normal form (CNF). A formula F is in CNF if and only if F is the conjunction of some clauses $c_1, \cdots, c_m$ in the form of $c_1 \wedge c_2 \wedge \cdots \wedge c_m$, where each clause is the disjunction of k literals $l_1, \cdots, l_k$, in the form of $l_1 \vee \cdots \vee l_k$. In this definition, each literal is either a variable x_i or negation of a variable $\overline{x_i}$.

For example, $(x_1 \vee \overline{x_2} \vee x_4) \wedge (\overline{x_1} \vee \overline{x_2} \vee x_3)$ is a satisfiability formula, and $x_1 = 1, x_2 = 0, x_3 = 1, x_4 = 1$ is a value-assignment to the variables satisfying the formula.

The defined satisfiability problem belongs to NP-complete class of complexity [1]. No polynomial solution for NP-complete problems is found yet, and nobody knows whether such a solution exists.

Without loss of generality, we assume that there is no clause containing both a variable and its negation, because if so, that clause is evaluated logically as "true" and could be omitted. We also assume that the literals in each clause are distinct. Hence, each clause contains at most n distinct literals, where n is the number of variables.

3   The Optical Satisfiability Solution

The main idea to solve the satisfiability problem is to generate all possible value-assignments for n variables and determine if there is at least one value-assignment satisfying the given formula or not. To do this, we show how to use light rays as value-assignments and generate all possible value-assignments. Then we show how to construct literal filters and clause filters to block those light rays that don't satisfy the given formula.

3.1   Light Rays as Value-Assignments

Consider a ribbon divided into 2^n equal cells, as there are exactly 2^n possible value-assignments for n variables (0 or 1 for each variable). We consider each value-assignment as a bit string, where the i-th position in the string is the value assigned to the i-th variable and assign each possible value-assignment to a cell in lexicographical order. An example of the value-assignment mapping for $n = 3$ is shown in Fig. 1. We say that a cell and light rays passing through it satisfy a given literal, clause, or formula if and only if the cell is assigned to a value-assignment satisfying the given literal, clause, or formula.

$$x_1 \quad x_2 \quad x_3$$

0	0	0
0	0	1
0	1	0
0	1	1
1	0	0
1	0	1
1	1	0
1	1	1

Fig. 1. Mapping the cells in a light ribbon to possible value-assignments for three variables

3.2   Literal Filters

A literal filter for literal l denoted by f_l is a ribbon having 2^n cells in such a form that a cell is transparent if and only if it satisfies l, and is opaque otherwise. We create $2n$ literal filters for literals $x_1, \cdots, x_n$ and $\overline{x_1}, \cdots, \overline{x_n}$. First we explain the structure of literal filters and then we describe how to create them efficiently.

Literal Filter Structure. To understand the shape of a literal filter f_{x_i} ($f_{\overline{x_i}}$), note that the value-assignments are mapped to the cells of a ribbon in lexico-graphical order, considering each value-assignment as a string where the i-th position in the string is the value assigned to x_i. So, each value-assignment for the first $i - 1$ variables appears in exactly $2 \times 2^{n-i}$ continuous cells. In these $2 \times 2^{n-i}$ cells, the first 2^{n-i} cells have value "0" for x_i so they are opaque in f_{x_i} (transparent in $f_{\overline{x_i}}$). The second 2^{n-i} cells have value "1" for x_i, so they are transparent in f_{x_i} (opaque in $f_{\overline{x_i}}$). Hence, the cells in f_{x_i} ($f_{\overline{x_i}}$) can be con-sidered as 2^i groups of cells, each group contains 2^{n-i} cells, and the groups are opaque and transparent (transparent and opaque) iteratively, as is shown in Fig 2. Literal filters f_{x_1}, f_{x_2}, and $f_{\overline{x_3}}$ are shown in Fig. 3.

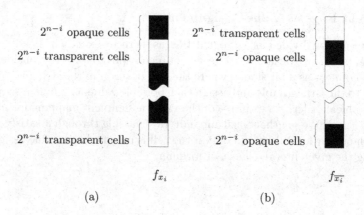

$$f_{x_i}$$

$$f_{\overline{x_i}}$$

(a)   (b)

Fig. 2. Each literal filter is consists of 2^i groups of 2^{n-i} cells, where the groups are opaque and transparent (transparent and opaque) iteratively (a) the literal filter f_{x_i} satisfying x_i, (b) the literal filter $f_{\overline{x_i}}$ satisfying $\overline{x_i}$

Creating Literal Filters. To create a literal filter, we use a ribbon shaped photographic film and divide it into 2^n cells. Each point in photographic film is transparent initially, and becomes opaque if light reaches it. By fixing the film chemically, the state of the cells (opaque or transparent) will not be changed any more. If light is emitted to a fixed film, transparent cells pass the light, and opaque cells block it.

To create literal filters f_{x_i} and $f_{\overline{x_i}}$, consider two ribbon shape photographic films denoted by r and $\overline{r}$ which are divided into 2^n cells and are transparent initially. At the first step, we emit light into the first 2^{n-i} cells of r and the

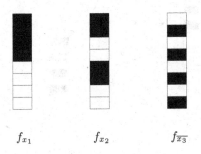

f_{x_1} f_{x_2} $f_{\overline{x_3}}$

Fig. 3. Examples of literal filters, f_{x_1}, f_{x_2}, and $f_{\overline{x_3}}$

second 2^{n-i} cells of $\overline{r}$ to become opaque. Then we fix the first $2 \times 2^{n-i}$ cells of r and $\overline{r}$, since these cells are already formed as they appear in f_{x_i} and $f_{\overline{x_i}}$ respectively.

Now suppose that in the k-th ($k = 1, 2, \cdots, i-1$) step, the first $2^k \times 2^{n-i}$ cells of r and $\overline{r}$ are shaped as they appear in f_{x_i} and $f_{\overline{x_i}}$ respectively. So, the first $2^k \times 2^{n-i}$ cells in r and $\overline{r}$ are in 2^k groups each of which containing 2^{n-i} cells, and the groups are opaque and transparent in r (transparent and opaque in $\overline{r}$) alternatively.

Now we place $\overline{r}$ in front of r in such a way that the first cell of $\overline{r}$ be placed in front of the $2^k \times 2^{n-i} + 1$-th cell of r (the first cell of r which is not fixed yet) and emit light to the first $2^k \times 2^{n-i}$ cells of $\overline{r}$ as is shown in Fig. 4a. Now the second $2^k \times 2^{n-i}$ cells of r are shaped as they appear in f_{x_i} (shown in Fig. 4b), and we fix them.

We do the same process for $\overline{r}$, by placing r in front of $\overline{r}$ where the first cell of r is placed in front of the $2^k \times 2^{n-i} + 1$-th cell of $\overline{r}$ and emit light to the first $2^k \times 2^{n-i}$ cells of r as is shown in Fig. 4c. Now, the second $2^k \times 2^{n-i}$ cells of $\overline{r}$ are also shaped as they appear in $f_{\overline{x_i}}$ (shown in Fig. 4d) and we fix them.

By applying the explained process, after k-th step, the first $2^{k+1} \times 2^{n-i}$ cells of r and $\overline{r}$ are fixed correctly.

After continuing this process for $i - 1$ steps $k = 1, 2, \cdots, i-1$, all 2^n cells of r and $\overline{r}$ are formed correctly and finally r is a literal filter for x_i and $\overline{r}$ is a literal filter for $\overline{x_i}$.

3.3   Clause Filters

A clause filter for clause $C = l_1 \vee \cdots \vee l_k$ denoted by f_C, is a ribbon divided into 2^n cells, where each cell is transparent if and only if it satisfies C, and is opaque otherwise.

Having $2n$ literal filters $f_{x_1}, f_{\overline{x_1}}, \cdots, f_{x_n}, f_{\overline{x_n}}$, we want to create m clause filters for m clauses $C_1, \cdots, C_m$ appeared in the given satisfiability formula. To create a clause filter f_C, note that a cell satisfies C if and only if it satisfies at least one of literals $l_1, \cdots, l_k$. Using the relation $\overline{C} = \overline{l_1} \wedge \cdots \wedge \overline{l_k}$, we place k literal filters $f_{\overline{l_1}}, \cdots, f_{\overline{l_k}}$ next to each other (the order is not important), then place a

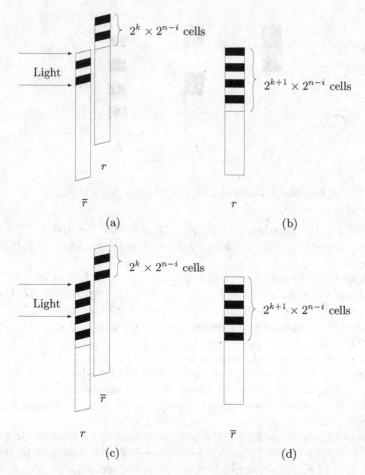

Fig. 4. The k-th step in creating literal filters f_{x_i} and $f_{\overline{x_i}}$. (a) Placing the first cell of $\overline{r}$ in front of the $2^k \times 2^{n-i} + 1$-th cell of r and emitting light. (b) The second $2^k \times 2^{n-i}$ cells of r are shaped as they appear in f_{x_i}. (c) Placing the first cell of r in front of the $2^k \times 2^{n-i} + 1$-th cell of $\overline{r}$ and emitting light. (d) The second $2^k \times 2^{n-i}$ cells of $\overline{r}$ are shaped as they appear in $f_{\overline{x_i}}$.

transparent ribbon shape photographic film r after them and emit light through the literal filters to reach r, as is shown in Fig. 5. Now, a cell in r is transparent if and only if it is opaque in at least one of $f_{\overline{l_1}}, \cdots, f_{\overline{l_k}}$. Note that a cell is opaque in at least one of $f_{\overline{l_1}}, \cdots, f_{\overline{l_k}}$ if and only if it is transparent in at least one of $f_{l_1}, \cdots, f_{l_k}$. Hence, the cells of r are formed as the cells appear in f_C. Then, we fix r, and f_C is created.

3.4 Finding the Answer

To find a given formula $F = C_1 \wedge \cdots \wedge C_m$ is satisfiable or not, we first create m clause filters $f_{C_1}, \cdots, f_{C_m}$. Note that F is satisfiable if and only if there exists at

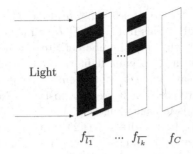

Fig. 5. Creating the clause filter f_C, where $C = l_1 \vee \cdots \vee l_k$

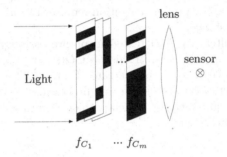

Fig. 6. Finding if the given formula $F = C_1 \wedge \cdots \wedge C_m$ is satisfiable or not

least one value-assignment satisfying F. In the other words, F is satisfiable if and only if there exists at least one cell which is transparent in all $f_{C_1}, \cdots, f_{C_m}$. We place m clause filters $f_{C_1}, \cdots, f_{C_m}$ next to each other (order is not important) and emit light from one side as is shown in Fig. 6. Using a lens and an optical sensor, we check if some light rays pass through the filters or not. If some light rays pass through all filters, there is at least one cell which is transparent in all filters, hence, F is satisfiable, F is not satisfiable otherwise.

4   Complexity

In this section, first we have investigated the length complexity of the required ribbon shaped photographic films and then, we have explained the complexity of time and energy consumed to solve each problem instance.

To solve a satisfiability problem instance having m clauses over n variables, each filter contains 2^n cells. Since $2n$ literal filters and m clause filters are created, total length of required photographic film is $O((m + n)2^n))$.

In the provided solution, the literal filters f_{x_i} and $f_{\overline{x_i}}$ are created in i steps and each step takes $O(1)$ time. Hence, $O(n^2)$ time is required to create all literal filters, for all n variables and their negations. After creating literal filters, m

clause filters are created, and f_{C_i} (for $0 < i \leq m$) takes $O(k_i)$ time to be created, where k_i is the number of literals in C_i. So, $O(K)$ time is required to create all clause filters, where K is the total number of literals in F, or $K = \sum_{i=1}^{m} k_i$. Finally, $O(1)$ time is required to emit light through m clause filters and find if the given formula is satisfiable or not. Hence, the provided solution takes $O(n^2 + K)$ time overall to solve each problem instance. Note that $K \leq mn$, as there are at most n literals in each clause.

In the phase of creating literal filters, $O(n2^n)$ light rays are required, because we have to create $2n$ literal filters each of which containing 2^n cells and each cell requires a light ray to be shaped. Similarly in the phase of creating clause filters, $O(m2^n)$ light rays are required because we have to create m clause filters each of which containing 2^n cells and each cell requires a light ray to be shaped. At the end, the answer of the problem is found by emitting $O(2^n)$ light rays to the clause filters. Thus, the provided solution requires $O((m+n)2^n)$ photons to solve each problem instance.

Note that literal filters can be created in preprocessing phase and be used to solve several problem instances. In this case, $O(n^2)$ time and $O(n2^n)$ photos are required in preprocessing phase to create all literal filters to solve problem instances over at most n variables. After preprocessing phase, $O(K)$ time and $O(m2^n)$ photos are required to solve each problem instance over n variables, m clauses, and K total literals.

5   Discussion

In the provided solution, each optical filter has exponential length but they are created in polynomial time by a novel method. The overall time to solve the problem is also polynomial which is the advantage of the provided solution in comparison to previous works based on delayed signals [12] and wavelength based approaches [9,10]. The provided solution is also practical for larger number of variables in comparison to wavelengths based approaches, as the spectrum space is divided into exponential number of zones in wavelength based approaches [9,10]. Note that although wavelength based approaches have exponential preprocessing time and limitation on large number of variables, but they use simpler optical filters and solve each problem instance in shorter time [9,10]. The other advantage of the provided solution is that there is no need to exponential number of optical devices in contrast to some previous works which split light rays exponential times [11].

Beside the advantages, the provided solution has practical limitations because of the exponential length of optical filters. Considering a standard $35mm$ photographic film resolving 6000 lines, 20 meters film is required to create each filter when $n = 40$ and 80 meters film is required to create each filter when $n = 43$. Note that, it is not necessary to emit light to the whole film ribbon simultaneously, or use a large lens. Instead, we can use a small light source and an optical sensor, and pass the film ribbon from the first to the end between them. It should be noticed that the number of photons required to each problem instance is exponentially increased according to number of variables.

6   Conclusion and Future Works

We have provided a novel polynomial time optical solution for the satisfiability problem. We have shown how to create exponential length optical filters in polynomial time, using photographic films. The filters are then used to specify light rays corresponding to proper answers of the satisfiability problem. The solution creates polynomial number of such filters to determine the answer of the given satisfiability problem. The number of photons used for each problem instance is exponential.

As the future work, we will try to apply the provided solution for other hard combinatorial problems. We will also try to use other properties of light, to reduce the length of filters is another issue, which brings the ability of solving problem instances over more variables. Another issue, is to generalize the optical operation used in this paper, to obtain a general optical computational model, which can be used to solve a wide range of problems.

References

1. Cormen, T.H., Leiserson, C.E., Rivest, R.L., Stein, C.: Introduction to Algorithms, 2nd edn. The MIT Press (2001)
2. Cook, S.A.: The complexity of theorem-proving procedures. In: Proceedings of the Third Annual ACM Symposium on Theory of Computing, pp. 151–158. ACM (1971)
3. Baier, C., Katoen, J.P., Larsen, K.G.: Principles of Model Checking. The MIT Press (2008)
4. Ganai, M., Gupta, A.: SAT-Based Scalable Formal Verification Solutions, 1st edn. Springer (2007)
5. Paturi, R., Pudlak, P.: On the complexity of circuit satisfiability. In: Proceedings of the 42nd ACM Symposium on Theory of Computing, pp. 241–250. ACM (2010)
6. Wen-Zhang, L., Jing-Fu, Z., Gui-Lu, L.: A parallel quantum algorithm for the satisfiability problem. Communication in Theoretical Physics 49(3), 629–630 (2008)
7. Johnson, C.R.: Automating the DNA computer: Solving n-variable 3-SAT problems. Natural Computing 7(2), 239–253 (2008)
8. Cecilia, J.M., García, J.M., Guerrero, G.D., del Amor, M.A.M., Pérez-Hurtado, I., Pérez-Jiménez, M.J.: Simulating a P system based efficient solution to SAT by using GPUs. The Journal of Logic and Algebraic Programming 79(6), 317–325 (2010)
9. Goliaei, S., Jalili, S.: An optical wavelength-based solution to the 3-SAT problem. In: Dolev, S., Oltean, M. (eds.) OSC 2009. LNCS, vol. 5882, pp. 77–85. Springer, Heidelberg (2009)
10. Goliaei, S., Jalili, S.: An optical solution to the 3-SAT problem using wavelength based selectors. International Journal of Supercomputing 62, 663–672 (2012)
11. Dolev, S., Fitoussi, H.: Masking traveling beams: Optical solutions for NP-complete problems, trading space for time. Theoretical Computer Science 411, 837–853 (2010)
12. Oltean, M., Muntean, O.: An optical solution for the SAT problem. In: Dolev, S., Oltean, M. (eds.) OSC 2010. LNCS, vol. 6748, pp. 53–62. Springer, Heidelberg (2011)

13. Head, T.: Parallel computing by xeroxing on transparencies. In: Condon, A., et al. (eds.) Algorithmic Bioprocesses. Natural Computing Series, pp. 631–637. Springer, Heidelberg (2009)
14. Goliaei, S., Jalili, S., Salimi, J.: Light-based solution for the dominating set problem. Applied Optics 51(29), 6979–6983 (2012)
15. Goliaei, S., Jalili, S.: Optical graph 3-colorability. In: Dolev, S., Oltean, M. (eds.) OSC 2010. LNCS, vol. 6748, pp. 16–22. Springer, Heidelberg (2011)

To What Extent Is Zero Energy Computing Feasible?

Joseph Shamir

Department of Electrical Engineering
Technion - Israel Institute of Technology, Haifa 32000, Israel
jsh@ee.technion.ac.il

Dedicated to the memory of H. John Caulfield

Abstract. Various data handling processes can be implemented without involving energy dissipation. However, these processes execute only part of a complete computing task and the remaining part will involve the loss of energy. This paper discusses some misleading concepts of reversible logic and presents a novel approach toward optical architectures with reduced energy consumption.

1   Introduction

About three decades of fruitful cooperation with H. John Caulfield had a sad and abrupt ending. This paper is the first attempt to proceed along the tracks we set out together, but with the regretful absence of John.

The essence of this paper can be summarized by the words of John Caulfield contributed to a joint paper which, unfortunately, was never completed: *"Everyone who owns a laptop computer knows that energy must be dissipated in computers. They get really hot. But we know that computers use logic gates in their construction, so seeking zero energy logic gates seems a good place to start reducing the required energy. In reality, zero energy logic gates can be made as shown here, but large computers cannot be made off these elements..."*

Traditional computing is executed by arrays of logic gates implementing Boolean logic. Since a fundamental logic operation has two inputs and only one output, information is lost on the way and the input cannot be reconstructed from the output. We say that this operation is irreversible and the loss of information is also associated with the loss of energy[1–3]. As these gates are intrinsically dissipative, there is a theoretical lower limit to the energy cost of each operation. From fundamental thermodynamic considerations this lower energy limit is $kT \ln 2$, where k is Boltzman's constant and T is the absolute temperature. This energy will heat our computers regardless of the technology adopted for implementing the logic gates. Needless to say that energy dissipation in state of the art computing technology is about five orders of magnitude above this theoretical energy dissipation limit.

To mitigate energy losses in logic gates several concepts of reversible gates were introduced during the years. The most frequently addressed such reversible

S. Dolev and M. Oltean (Eds.): OSC 2012, LNCS 7715, pp. 25–34, 2013.
© Springer-Verlag Berlin Heidelberg 2013

gate is the Fredkin gate[4]. It turns out that these gates are ideal for optical implementation[5] and, under certain conditions, they can operate with no energy dissipation. The next section addresses the basic concepts of reversible logic and this is followed by a demonstration that specific logic functions can be implemented on a generic interconnection network.

2   Reversible Logic – General Considerations

As indicated in the introduction, conventional logic gates discard information and, as a consequence, are dissipative in terms of information and energy[1]. In contrast, reversible logic is based on lossless logic elements, such as the Fredkin gate[4]. Essentially, a Fredkin gate is a cross-bar switch converted to perform logic operations. The schematic representation of a Fredkin gate is depicted in Fig. 1. Traditionally, a Fredkin gate has three information channels, the control channel C and two data input and output channels. While the signal in the control channel is transmitted unchanged it controls the two data channels, in the following way:

$$b_1 = a_1; \quad b_2 = a_2 \quad \text{If} \quad C = 0; \quad b_1 = a_2; \quad b_2 = a_1 \quad \text{If} \quad C = 1. \tag{1}$$

It is appropriate to note that this definition is the complement of the original definition by Fredkin and Toffoli [4] since we found the present definition more convenient for our applications. The Fredkin gate is lossless and reversible in the sense that if all three outputs are detected, the input can be reconstructed. The energy needed to change the state of the gate between $C = 0$ and $C = 1$ is ignored in most publications.

The controlled switching operation of the Fredkin gate can be represented mathematically in several ways[3] out of which we found it convenient to employ here the matrix representation given as,

$$\begin{pmatrix} b_1 \\ b_2 \end{pmatrix} = M \begin{pmatrix} a_1 \\ a_2 \end{pmatrix} \tag{2}$$

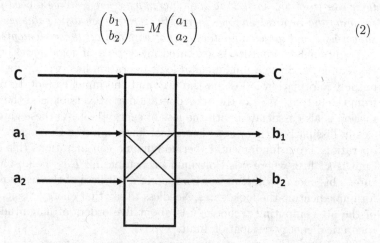

Fig. 1. Basic definition of the Fredkin gate

where,

$$M = \begin{pmatrix} 1 & 0 \\ 0 & 1 \end{pmatrix} \text{ for } C = 0; \qquad M = \begin{pmatrix} 0 & 1 \\ 1 & 0 \end{pmatrix} \text{ for } C = 1. \qquad (3)$$

The original idea of Fredkin gates was to use the two input channels for the data and the control channel, C, for controlling the gate's state. However, already from the start, the proposal to use these reversible gates for the implementation of traditional Boolean logic functions[3–5] mixed up the role of the data and control channels. In principle, Boolean logic functions can be performed with one, or a combination of several reversible elements as shown in Fig. 2 for the three primitive logic elements, NOT, AND and OR gates. Looking at the architectures of these implementations we observe that each logic operation is accompanied by two additional outputs filling up all the three output channels of the gate. In the terminology of reversible logic the outputs irrelevant to the specific operation are considered *garbage*. Unfortunately, if this garbage is discarded reversibility is lost and energy dissipation can no longer be avoided. In other words we may state that any "black box" that performs the function of a conventional logic gate, for all external aspects it is a logic gate and an external observer cannot distinguish between them. This means that the operation of this black box will be associated, among other characteristics, with losses regardless of its internal

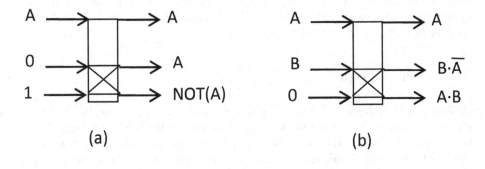

(a)   (b)

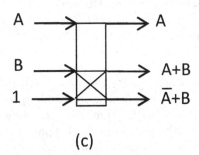

(c)

Fig. 2. Implementation of logic elements using a Fredkin gate. (a) NOT gate, (b) AND gate and (c) OR gate.

constitution, even if each internal component is nominally lossless. With these considerations we may conclude that traditional ways to implement Boolean logic operations with reversible logic gates reinstate dissipation of at least the same order of magnitude as conventional logic gates. Thus, the whole idea of reversible computing is contradicted.

Another difficulty encountered in the implementation of Boolean logic operations using several Fredkin gates is the mixing of the control and data channels. The problem stems from the fact that the physical nature of the control channel is usually completely different than the two data channels. In practice, while the original data channels flow with no theoretical lower limit on energy dissipation, the control channel must use energy, at least for changing the state of the gate. Moreover, to maintain a certain state, the switching energy between the two states must significantly exceed the thermal noise of the order of kT. In practice this energy will be much higher than the $kT \ln 2$ theoretical limit[1, 3] on a single logic operation.

As a consequence of the above considerations we have to conclude that lossless computing must involve modifications of the whole computing paradigm. Perhaps the most promising rout to go is the concept of hybridization where the computing architecture contains reversible components as well as non-reversible ones. A computing architecture based on this concept will perform part of the process with no energy loss but the rest will still involve energy dissipation. An interesting approach exploiting this concept is based on the recently introduced concept of Directed Logic[6] (DL).

3 Directed Logic

Basically, a DL network [6] comprises a combination of interconnected Fredkin gates where input data is fed to the control channel of the gates and logic functions are implemented by light propagating through the data channels. Thus, the need to mix signals of different physical nature is eliminated and information in the data channels flows along the net with no obstruction and no theoretical lower limit on energy dissipation.

An apparent disadvantage of the DL approach as compared to traditional Fredkin gate logic operations is a loss of the idealized simplicity indicated in Fig. 2. For example, to implement an OR gate within the DL approach we need three gate elements (Fig. 3) instead of just one and one of the input signals must be fed to two elements in parallel. However, in view of the above discussion it is obvious that there is an actual penalty paid for the simplicity of the original Fredkin gate in terms of practical system complexity, processing speed and energy dissipation. Energy dissipation is particularly large when the state of the gate is switched. While in the original Fredkin gate implementations the gates are switched during the input and the processing operation, in DL the gate is switched only with the input data.

With all the indicated advantages of DL it still suffers from inflexibility since the network must be rewired if the required logic function is changed.

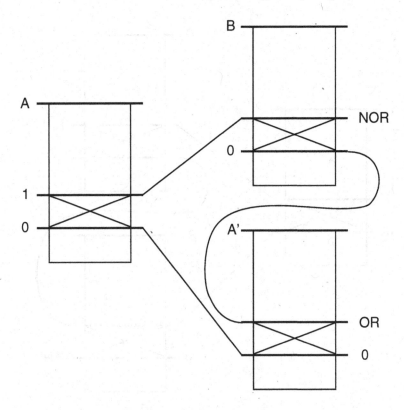

Fig. 3. DL implementation of an OR gate

This inflexibility can be represented by the rewiring needed on the network of Fig. 3 to implement other logic functions, such as an AND operation as shown in Fig. 4.

To mitigate the inflexibility of DL networks, in the next section we introduce a generic architecture into which different logic operations can be programmed in a flexible way.

4   Programmable Directed Logic Network

As indicated above and in the analysis provided in Ref. [6] the advantages of DL significantly out weight its disadvantages and the feasibility of the practical optical implementation of DL elements and small networks has been already demonstrated in several laboratory experiments.[7–11] For our purpose here we shall represent the basic building block of the network by the optical waveguide coupler, schematically shown in Fig. 5. Such a coupler consists of two adjacent optical waveguides with an intersection, represented by the ellipse, where the coupling between them can be controlled. There are many possible implementations for such a coupler, including the Mach-Zehnder coupler, which is already

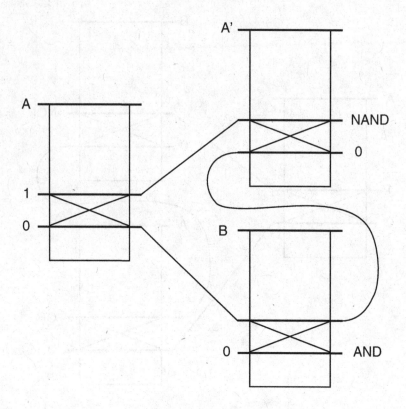

Fig. 4. DL implementation AND and NAND functions

widely used in communication networks, micro-ring resonators, electro-optically activated couplers, thermal couplers and wavelength sensitive couplers.

The optical waveguide coupler was already singled out in Ref. [5] as a useful implementation of an elementary Fredkin gate. Since then several architectures were proposed where the waveguide coupler also served as a building block for the implementation of interconnection networks in two [12] and three-dimensional architectures[13] as well as arithmetic processors[14]. In this paper we shall also consider the optical waveguide coupler implementation of the Fredkin gate and, accordingly, from now on the traditional data channels will be referred to as light channels. This terminology is useful since in the DL approach there is no interchange between light channels and control channels, keeping the light channels for nominally lossless information transfer. The dissipative input interface is combined with the control channel eliminating the need to transfer the data from a separate input device to the processor.

A generic interconnection network composed of waveguide couplers is shown in Fig. 6. A complete permutation network between $2N$ input channels and $2N$ output channels can be implemented by $2N$ *switching layers*, where such a layer contains all the gates along a vertical line [12] and a total of $N(2N-1)$ switching

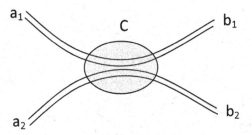

Fig. 5. A waveguide coupler as Fredkin gate. The two waveguide channels are coupled by a controllable element represented by the ellipse.

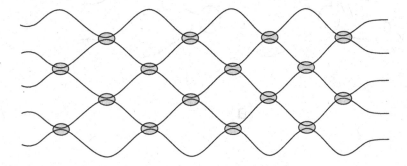

Fig. 6. A generic interconnection network

elements. In the implementation of a DL processor based on such an interconnection network we do not need all possible permutations but each computing gate element has only one data input port (the control input). Therefore we need at least as many gates as we have input data bits. As a consequence, to implement a usual two-input logic gate we shall need at least two DL gates and, in general, to process N data inputs we shall need at least N DL gates. As we demonstrate here the generic permutation interconnection architecture of Fig. 6, which can direct in parallel any of the $2N$ light input ports to any of the $2N$ light output ports, can be exploited for logic operations. Within the original DL paradigm this means that we supply a light input vector with only one non-zero element and the result of a logic operation is detected on one of the output ports. While this already is a significant improvement over the original hard-wired DL network here we allow several non-zero input elements to perform several logic operations in parallel. For the implementation of each logic function described below we employ a limited section of the complete interconnection net, allowing for additional operations to be performed in parallel.

To demonstrate the improved flexibility of the programmable array we present here the implementation of the OR and AND operations of Figs. 3 and 4. Fig. 7 shows a section of an interconnection network of four channels and six gates, not all active for this implementation. The active gates are marked by the input data

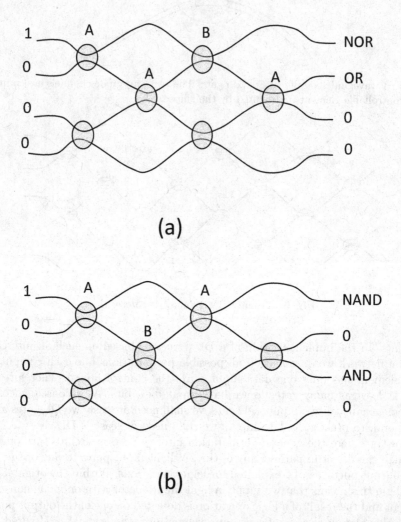

(a)

(b)

Fig. 7. Section of interconnection network implementing an OR gate (a) and an AND gate (b). See text for more detail.

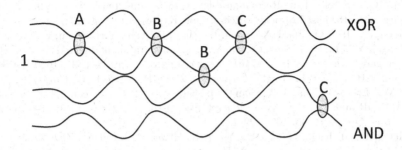

Fig. 8. Implementation of three-input XOR and AND operations in parallel with a single finite light input in the second channel from top

A and B. To implement an OR operation (Fig. 7a), there is one light channel with a finite input, the A input is fed to three gates simultaneously while the B input is fed to one gate. To implement the AND operation it is adequate to feed the A and B to one gate each. To obtain also the NAND operation the A input is fed also to the second A gate to the right. Taking a longer section, we can implement a three-input XOR and AND operation simultaneously as show in Fig. 8. It is straight forward to extend this architecture to multiple inputs and more complicated logic functions can be also implemented, sometimes with several finite inputs to the light channels.

5 Conclusions

Reversible logic elements are not adequate to perform full computing operations. Nevertheless, discrete logic functions can be executed with no lower bound of energy dissipation. Integrating such logic operations in general computing architectures can lead to significantly reduced energy dissipation, high operating speed and flexibility.

References

1. Landauer, R.: Irreversibility and heat generation in the computing process. IBM Journal of Research and Development 5, 183–191 (1961)
2. Bennett, C.H.: Logical reversibility of computation. IBM Journal of Research and Development 17, 525–532 (1973)
3. Feynman, R.P.: Quantum mechanical computing. Optics News 11, 11–20 (1985)
4. Fredkin, E., Toffoli, T.: Conservative Logic. Int. J. Theoret. Phys. 21, 219–253 (1982)
5. Shamir, J., Caulfield, H.J., Miceli, W., Seymour, R.J.: Optical Computing and the Fredkin Gate. Appl. Opt. 25, 1604–1607 (1986)
6. Hardy, J., Shamir, J.: Optics inspired logic architecture. Optics Express 15(1), 150–165 (2007)

7. Xu, Q., Soref, R.: Reconfigurable optical directed-logic circuits using microresonator-based optical switches. Opt. Express 19, 5241 (2011)
8. Zhang, L., Ji, R.Q., Jia, L.X., Yang, L., Zhou, P., Tian, Y.H., Chen, P., Lu, Y.Y., Jiang, Z.Y., Liu, Y.L., Fang, Q., Yu, M.B.: Demonstration of directed XOR/XNOR logic gates using two cascaded microring resonators. Opt. Lett. 35, 1620 (2010)
9. Tian, Y.H., Zhang, L., Ji, R.Q., Yang, L., Zhou, P., Chen, H.T., Ding, J.F., Zhu, W.W., Lu, Y.Y., Jia, L.X., Fang, Q., Yu, M.B.: Proof of concept of directed OR/NOR and AND/NAND logic circuit consisting of two parallel microring resonators. Opt. Lett. 36, 1650 (2011)
10. Zhang, L., Ji, R.Q., Tian, Y.H., Yang, L., Zhou, P., Lu, Y.Y., Zhu, W.W., Liu, Y.L., Jia, X.L., Fang, Q., Yu, M.B.: Simultaneous implementation of XOR and XNOR operations using a directed logic circuit based on two microring resonators. Opt. Express 19, 6524 (2011)
11. Papaioannou, S., Vyrsokinos, K., Tsilipakos, O., Pitilakis, A., Hassan, K., Weeber, J.-C., Markey, L., Dereux, A., Bozhevolnyi, S.I., Miliou, A., Kriezis, E.E., Pleros, N.: A 320 Gb/s-Throughput Capable 2x2 Silicon-Plasmonic Router Architecture for Optical Interconnects. J. Lightwave Tech. 29(21) (November 2011)
12. Shamir, J., Caulfield, H.J.: High-efficiency rapidly programmable optical interconnections. Appl. Opt. 26, 1032–1037 (1987)
13. Shamir, J.: Three-dimensional optical interconnection gate array. Appl. Opt. 26, 3455–3457 (1987)
14. Mirsalehi, M.M., Shamir, J., Caulfield, H.J.: Residue arithmetic processing utilizing optical Fredkin gate arrays. Appl. Opt. 26, 3940–3946 (1987)

All-Optical XOR Gate for QPSK In-Phase and Quadrature Components Based on Periodically Poled Lithium Niobate Waveguide for Photonic Coding and Error Detection Applications

Emma Lazzeri[1], Antonio Malacarne[2], Giovanni Serafino[1], and Antonella Bogoni[2]

[1] TeCIP, Scuola Superiore Sant'Anna, Pisa, Italy
emma.lazzeri@cnit.it
[2] CNIT, Pisa, Italy

Abstract. An all-optical scheme based on periodically poled lithium Niobate waveguide (PPLN) for signal processing of the in-phase (I) and quadrature (Q) components of an input quadrature phase shift keying (QPSK) signal is presented. The device is able to work on the I and Q components without any additional demodulation stage and makes use of cascaded second harmonic and difference frequency generation (SHG and DFG respectively) in the PPLN waveguide to obtain the logical operation XOR(I,Q). A single continuous wave pump signal is needed in addition to the input signal to generate the output signal, in which the information is coded in a binary phase shift keying (BPSK) modulation. The logical XOR(I,Q) is a basic operation that can enable data coding and error detection in all-optical networks. Bit error rate measurements are provided to evaluate the system performance for a 20Gb/s DQPSK input signal, and tunability of the output signal wavelength has been attested with almost constant optical signal-to-noise-ratio (OSNR) penalty along the C-band.

Keywords: Periodically Poled Lithium Niobate Waveguide, Quadrature Phase Shift Keying, All-Optical Processing, XOR, Coding, Error Detection.

1 Introduction

The increasing demand of bandwidth arising from new and more complex telecom services and applications is spurring the research towards the study of spectral efficient transmission systems and the design of devices able to overcome the actual electronic bottleneck of the existing telecommunication networks.

Phase shift keying (PSK) modulation formats are attractive due to their low bandwidth occupation and their high tolerance to chromatic dispersion that allows a more efficient allocation of limited resources, and a longer reach compared to the transmission systems based on amplitude phase shift keying (ASK) modulation. Recently quadrature phase shift keying (QPSK) has started to be exploited in the long-haul segment for existing optical networks and its use is expected to be extended to the metro and access segments to improve the networks capacity.

S. Dolev and M. Oltean (Eds.): OSC 2012, LNCS 7715, pp. 35–41, 2013.
© Springer-Verlag Berlin Heidelberg 2013

All-optical processing techniques are in general desirable solutions to overcome the bottleneck of the existing telecommunication networks that still make a large use of electronic devices to process and regenerate the transmitted optical signals, and thus limit the tributary channel speed to tens of Gbps. Furthermore, optics has the potential to lower the power consumption with respect to current electronic solutions. The use of photonics technology applied to the logical computing of the optical signals has been proved to work at ultrahigh speed, up to Tbps [1]. A particularly interesting branch of the signal processing technology is the one concerning the coding and error detection, that could allow the telecommunication systems based on ultrafast optical technology to be more efficient and reliable. Simple examples of optical coding (OC) and error detection (OED) can arise from elementary logical operations such as XOR logic gates, and have been recently implemented [2], [3].

In this scenario, elementary XOR logic operation on the in-phase (I) and quadrature (Q) components of an ultra-fast all-optical QPSK signal, without the need of demodulating the original data stream to act on the single I and Q components, could allow OC and OED techniques such as checksum and cyclic redundancy check, and enhance the overall transmission system reliability and efficiency.

Several solutions have been proposed to perform XOR operation of ASK and PSK signals in the optical domain exploiting $\chi^{(3)}$ effects such as four wave mixing (FWM) in semiconductor optical amplifier (SOA) [4-5], and in highly nonlinear fiber (HNLF) [6-7], or cascaded $\chi^{(2)}$ processes in periodically poled lithium Niobate waveguide (PPLN) [8-9].

Among the mentioned techniques, cascaded $\chi^{(2)}$ processes in PPLN, such as second harmonic and difference frequency generation (SHG and DFG), have several advantages over the use of $\chi^{(3)}$ effects in SOA or HNLF. The transparency to data rate and modulation format, the absence of undesired $\chi^{(3)}$ effects, the broad conversion bandwidth allowing simultaneous processing of several signals, the negligible introduction of quantum-limited amplified spontaneous emission (ASE) noise, and low signal distortion make PPLN an attractive solution to perform processing of advanced modulation format signals. Moreover, the possibility to further increase the potential of PPLN in terms of conversion efficiency [10] and polarization insensitivity [11] has been investigated and demonstrated.

In this paper, the architecture of a simple and flexible device performing the XOR operation between the I and Q components of an input QPSK signal is presented, and an experimental validation of the operation principle acting on a 20Gbps QPSK input data stream is provided by means of bit error rate (BER) measurements. The device is based on cascaded SHG/DFG in PPLN and it is potentially polarization independent. Unlike the only one scheme already existing in literature [9], here the XOR between the I and Q components of the input QPSK signal is obtained without any particular condition on the input signal data stream. Moreover, the input signal central wavelength do not have to satisfy any particular condition, because by a change in the PPLN driving temperature, the scheme can be adapted to any input signal configuration, making the device colorless and flexible.

This paper is structured as follows. After the introduction giving an overview about the state of the art in optical coding to perform the XOR operation, a section describing the operation principle and the theoretical scenario is presented. Afterwards, the third section explains the employed setup and gives some details about the experimental activity, whereas in the fourth section the obtained results are discussed. Finally, in the Conclusions the relevant results are summarized.

2   Operation Principle

The operation principle of the all-optical device performing the logical XOR on the I and Q components of the input QPSK signal S_i is depicted in Fig. 1(left).

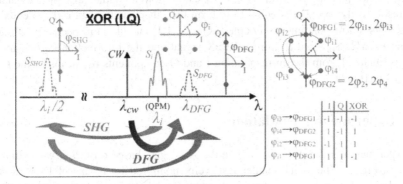

Fig. 1. Left: operating principle of the PPLN based XOR; Right: mapping of the input I and Q components on the output DPSK signal

The logic gate is based on cascaded SHG and DFG in PPLN waveguide. The PPLN temperature is set to make S_i central wavelength (λ_i) fall inside the quasi-phase matching (QPM) condition bandwidth. This way, when travelling through the PPLN, the input signal at wavelength λ_i experiences SHG and a new signal S_{SHG} is created at $\lambda_{SHG} = \lambda_i/2$. This signal carries the information about the XOR(I,Q) and is brought back to the C band by means of DFG between S_{SHG} and a continuous wave pump signal S_{CW} at λ_{CW}. The ultimate output of the device is therefore signal S_{DFG}, whose central wavelength λ_{DFG} satisfies the condition:

$$\frac{1}{\lambda_{DFG}} = \frac{1}{\lambda_{SHG}} - \frac{1}{\lambda_{CW}} \tag{1}$$

Fig. 1(right) shows how the information about the logical XOR(I,Q) is mapped into signal S_{SHG}. During the SHG process, two photons at λ_i are combined to generate a third photon at $\lambda_i/2$. Since signal S_i is a QPSK signal, the phase of Si corresponding to the k-th bit ($\varphi_i(k)$) is mapped into a new PSK signal S_{SHG} whose phase $\varphi_{SHG}(k)$ is given by:

$$\varphi_{DFG}(k) = \arg\{S_{DFG}(kT)\} = 2 \cdot \arg\{S_i(kT)\} = 2 \cdot \varphi_i(k) \tag{2}$$

where T is the bit interval and arg{S} is the argument of signal S.

The information carried by the output signal S_{DFG} is equal to the one of signal S_{SHG} as its phase $\varphi_{DFG}(k)$ is given by:

$$\varphi_{DFG}(k) = arg\{S_{DFG}(kT)\}$$
$$= arg\{S_{SHG}(kT)\} - arg\{S_{CW}(kT)\} = \varphi_{SHG}(k) - \varphi_{CW}(k) \tag{3}$$

where $\varphi_{CW}(k)$ is a constant value that we can set to be equal to 0 to simplify the problem description.

When $\varphi_i(k)$ is equal to $\pi/4$ or $5\pi/4$, corresponding to a couple (I,Q) of (1,1) and (-1,-1) respectively, $\varphi_{DFG}(k)$ is equal to $\pi/2$; on the other hand, when $\varphi_i(k)$ is equal to $3\pi/4$ or $7\pi/4$, corresponding to a couple (I,Q) of (-1,1) and (1,-1) respectively, $\varphi_{DFG}(k)$ is equal to $-\pi/2$. S_{DFG} is therefore a BPSK signal and the information carried by its phase is obtained from the corresponding I and Q components by means of the truth table of Fig.1(right).

3   Experimental Validation

The experimental setup used to validate the XOR gate based on PPLN waveguide is shown in Fig. 2. The signals under test were monitored by means of DQPSK and DPSK demodulators, but the scheme described can be used to operate also on QPSK signals. The setup consists in three blocks: transmitter, receiver and device under test, performing the XOR operation by means of cascaded SHG and DFG in PPLN as described in the previous paragraph.

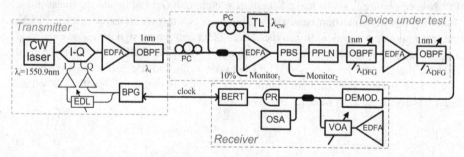

Fig. 2. Experimental setup

In the transmitter block, a DQPSK signal (S_i) is created by means of a continuous wave lasers, an I-Q modulator and a bit pattern generator (BPG). A fixed wavelength is modulated to create the DQPSK signal at $\lambda_i = 1550,9nm$. Two electrical drivers amplify the 10Gbps 2^7-1 pseudo-random bit sequence (PRBS) D and its delayed replica that drive the RF ports (I and Q) of the modulator, to produce a DQPSK signal at 10Gbaud/s (20Gbps). RF cables with different lengths and an electrical delay line (EDL) make sure that the driving sequences are uncorrelated. A low noise erbium doped fiber amplifier (EDFA) is used to amplify the DQPSK signal and 1nm wide optical band pass filter removes the out of band amplified spontaneous emission noise (ASE).

The pump signal is generated inside the device that performs the XOR operation by means of a tunable laser (TL). Two polarization controllers (PCs) are used to align the polarization of the pump to the one of S_i, and a 3dB coupler is used to combine the signals before entering a booster EDFA. At the EDFA input, the CW pump power level is around 3dBm, whereas the signal is at 0dBm. The EDFA saturated output power is about 28dBm, which means that, at the PPLN input, the two signals power level is around 25dBm. A polarization beam splitter (PBS) is used to polarize the signals that are fed into the polarization maintaining input pigtail of the waveguide. Finally, a tunable optical OBPF with a bandwidth of 1nm selects the output signal at λ_{DFG}, whose power level is 1.5dBm. Before entering an EDFA, a second OBPF (1nm wide) removes the exceeding ASE noise. The spectra at the input (a) and output (b) of the PPLN are shown in Fig. 3; in this case λ_{CW} = 1545.3nm and, consequently, λ_{DFG} = 1556.7nm. The Monitor$_1$ shown in Fig. 2 has been used for monitoring the average power (~1.5mW for each signal) of each signal at the booster EDFA input. For matching the right polarization state of each signal, thus maximizing their power, at the PPLN input, the power monitored at Monitor$_2$ has been minimized by properly adjusting each signal polarization state. The temperature of the PPLN has been kept at 19.1°C, to match the QPM condition to the input signal wavelength.

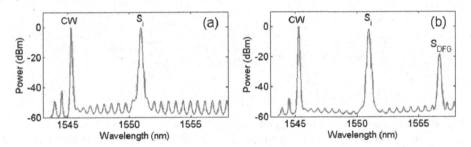

Fig. 3. PPLN input (a) and output (b) spectra

The receiver shown in Fig. 2 was used to perform BER measurements on the input and output signals. Noise loading is used to change the optical signal to noise ratio (OSNR) at the receiver section. The signal under test is first demodulated and then coupled to the noise coming from an EDFA. A variable optical attenuator (VOA) changes the amount of noise to vary the OSRN that is monitored by means of an optical spectrum analyzer (OSA). A 10GHz photoreceiver (PR) is used to convert the optical signal to an electrical data stream before entering the BER tester.

4   Results and Discussion

Fig 4(a) shows the BER vs OSNR for the input DQPSK I and Q components and for the output DPSK signal. The eye diagrams corresponding to the demodulated S_{DFG} and the I and Q components of the input signal are also shown in Fig. 4(c)-(e). Since the same 2^7-1 PRBS drives both the I and Q arms of the modulator, thanks to one of the fundamental properties of PRBS sequences, the resulting S_{DFG} is again the same PRBS, as it is the results of XOR(I,Q). Again, the ASK sequence at the output of the DPSK demodulator is the same PRBS. However, since no differential coder was available at the patter generator, the demodulated I and Q sequence at the output of the DQPSK demodulator are not the same PRBS, therefore it is not possible to show the I and Q sequences that produce the S_{DGF} demodulated output. Since a balanced photoreceiver was not available, a single photoreceiver was used to perform the measurements on the two arms of the DPSK demodulator output and on I_1, I_2, Q_1, and Q_2 outputs of the DQPSK demodulator respectively. A Power penalty of 4dB was measured for BER=10^{-9} (Fig. 4(a)).

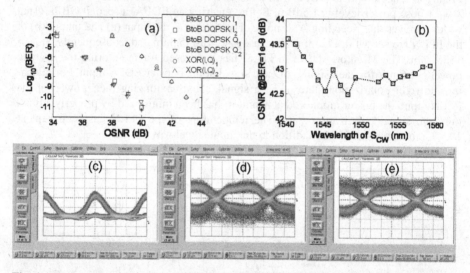

Fig. 4. BER measurements for the demodulated input and output signals (a). OSNR at the receiver section to obtain BER = 10^{-9} versus λ_{CW} (b). Eye diagram of I_1 component for the input DQPSK signal (c). Eye diagram of the DPSK demodulator output 1 (d) and output 2 (e) when the output signal of the system (XOR function) is applied.

To demonstrate the possibility to tune the output signal central wavelength and the independency of the device performance with respect to such wavelength, BER measurements were recorded while changing the pump signal (TL) central wavelength λ_{CW}. A graph reporting the value of OSNR needed to obtain a BER = 10^{-9} versus S_{CW} central wavelength is depicted in Fig. 4(b). All the OSNR values are contained within 1.1dB and this confirms the independence of the scheme performance on the pump wavelength.

5   Conclusions

We presented a new scheme performing the logical XOR of the in-phase and quadrature components of an input QPSK signal based on cascaded SHG and DFG in PPLN waveguide. The scheme does not require any additional demodulating or wavelength converting stage as by changing the PPLN temperature, the QPM condition can be adjusted to operate on the input signal central wavelength. All-optical XOR gates can enable optical coding and error detection in next generation optical networks. BER measurements on the input and output signals of the XOR gate show a power penalty of 4dB at BER=10^{-9}. Finally, tunability of the output signal wavelength is demonstrated together with OSNR penalty along the C-band comprised in 1.1dB.

Aknowledgments. This work has be funded by the Italian project ARNO.

References

1. Oxenløwe, L.K., Ji, H., Galili, M., Pu, M., Hu, H., Mulvad, H.C.H., Yvind, K., Hvam, J.M., Clausen, A.T., Jeppesen, P.: Silicon Photonics for Signal Processing of Tbit/s Serial Data Signals. IEEE Journal of Selected Topics in Quantum Electronics 18(2), 996–1005 (2012)
2. Wang, Z., Huang, Y.-K., Deng, Y., Chang, J., Prucnal, P.R.: Optical Encryption With OCDMA Code Swapping Using All-Optical XOR Logic Gate. IEEE Photonics Technology Letters 21(7), 411–413 (2009)
3. Aikawa, Y., Shimizu, S., Uenohara, H.: Demonstration of All-Optical Divider Circuit Using SOA-MZI-Type XOR Gate and Feedback Loop for Forward Error Detection. Journal of Lightwave Technology 29(15), 2259–2266 (2011)
4. Porzi, C., Scaffardi, M., Potì, L., Bogoni, A.: Optical Digital Signal Processing in a Single SOA Without Assist Probe Light. IEEE Journal of Selected Topics in Quantum Electronics 16(5), 1469–1475 (2010)
5. Chan, K., Chan, C.K., Chen, L.K., Tong, F.: Demonstration of 20-Gb/s all-optical XOR gate by fourwavemixing in semiconductor optical amplifier with RZ-DPSK modulated inputs. IEEE Photon. Technol. Lett. 16, 897–899 (2004)
6. Yu, C., Christen, L., Luo, T., Wang, Y., Pan, Z., Yan, L.-S., Willner, A.E.: All-optical XOR gate using polarization rotation in single highly nonlinear fiber. IEEE Photonics Technology Letters 17(6), 1232–1234 (2005)
7. Wang, J., Sun, Q., Sun, J., Zhang, X.: Experimental demonstration on 40 Gbit/s all-optical multicasting logic XOR gate for NRZ-DPSK signals using four-wave mixing in highly nonlinear fiber. Optics Communications 282(13) (July 2009)
8. Wang, J., Sun, J.: All-optical logic XOR gate for high-speed CSRZ-DPSK signals based on cSFG/DFG in PPLN waveguide. Electronics Letters 46(4), 288–290 (2010)
9. Kong, D., Li, Y., Wang, H., Zhang, X., Zhang, J., Wu, J., Lin, J.: All-Optical XOR Gates for QPSK Signals Based on Four-Wave Mixing in a Semiconductor Optical Amplifier. IEEE Photonics Technology Letters 24(12), 988–990 (2012)
10. Umeki, T., Asobe, M., Nishida, Y., Tadanaga, O., Magari, K., Yanagawa, T., Suzuki, H.: Highly efficient +5-dB parametric gain conversion using direct-bonded PPZnLN ridge waveguide. IEEE Photon. Technol. Lett. 20(1), 15–17 (2008)
11. Hu, H., Suche, H., Ludwig, R., Huettl, B., Schmidt-Langhorst, C., Nouroozi, R., Sohler, W., Schubert, C.: Polarization insensitive all-optical wavelength conversion of 320 Gb/s RZ-DQPSK data signals. In: Proc. OFC, March 22-26 (2009)

All-Optical Ultrafast Adder/Subtractor and MUX/DEMUX Circuits with Silicon Microring Resonators

Purnima Sethi and Sukhdev Roy[*]

Department of Physics and Computer Science
Dayalbagh Educational Institute, Dayalbagh, Agra 282 110 India
sukhdevroy@dei.ac.in

Abstract. We present designs of all-optical ultrafast simultaneous NOR logic gate/Half-Adder/Subtractor, Full-Adder/Subtractor and Multiplexer/De-Multiplexer circuits using add-drop silicon microring resonators. The proposed circuits require less number of switches and inputs for realization of the desired logic compared to earlier reported designs. Multiplexer/De-Multiplexer operations can be realized from the same circuit by simply interchanging the inputs and outputs. Lower energy consumption and delays along with reconfigurability and compactness make them attractive for practical applications.

Keywords: all-optical switching, optical computing, microring resonator, directed logic, silicon photonics.

1   Introduction

There is tremendous research effort to achieve all-optical information processing for ultrafast, ultrahigh bandwidth communication and computing. Advances in the fabrication of micro and nanostructures have opened up exciting possibilities to generate, modulate and detect light to achieve energy efficient optical supercomputing. Power optimization along with faster computing is a major technological challenge that requires new computing paradigms [1-4]. Directed Logic and Binary Decision Diagram are newly proposed strategies which minimize the latency in calculating a complicated logic function by taking advantage of fast and low-loss propagation of light in a highly integrated on-chip photonic system [4-7].

Silicon ring resonators provide a very versatile platform for optical switching and computing offering advantages of high-Q, high switching speed, low-power consumption, ease of fabrication and large scale integration [4-10]. Recently, silicon resonator operation over very wide temperature range (>80°C) has been demonstrated, rendering silicon photonic devices CMOS compatible [11]. Modulation with ring resonators is advantageous as they are compact and can be actuated directly as a lumped element, even at high speeds (10–25GHz). Also, the relatively small area limits the necessary power to modulate, making them hold the best potential in terms of modulation energy per bit [7,8]. To increase the bit rate and speed of operation, the free-carrier dynamics during injection and extraction can be controlled inside the optical device by a combination of several distinct mechanisms [12-13].

---

[*] Corresponding author.

S. Dolev and M. Oltean (Eds.): OSC 2012, LNCS 7715, pp. 42–53, 2013.

Resonator based optical switches modulated by electrical [5], thermal [6], or optical [14-17] signals have been widely investigated. Recently, an experimental demonstration of fast all-optical switching using silicon microring resonators has been shown [15, 16]. The transmission of the structure can be modulated by up to 94% ($\Delta\lambda = -0.36$ nm and $\Delta n = -4.8 \times 10^{-4}$) in less than 500 ps using light pulses with energies as low as 25 pJ [15].

High speed all-optical computing devices are building blocks for next-generation optical networks and computing systems as they are key elements in all-optical signal processing and future all-optical time-division multiplexed (OTDM) networks [1,2]. All-optical AND/NAND logic operations have been experimentally demonstrated at 310 Mbit/s with ~10-dB extinction ratio using a 1 x 1 silicon microring resonator [17]. Lin *et al.* have used 15 optical switches based on 1 x 2 silicon ring resonators to demonstrate a basic half-adder [7]. Xu and Soref have used reconfigurable optical directed logic architecture to realize full adder, MUX/DEMUX, encoder and comparator circuits by rewiring arrays of hundreds of 1 x 1 switches controlled by the logic input signals [4]. Their architecture offers significant improvements in reconfigurability and scalability. However, the above designs require large number of switches for realization of the desired logic. Reduction in the number of switches can substantially reduce the power consumption, delay, latency and size of the devices.

In our previous work, we have presented designs of all-optical Boolean as well as conservative and reversible logic gates with optically controlled microresonators [18,19]. Higher computing circuits such as half/full adder-subtractor, multiplexer/demultiplexer, and arithmetic logic unit circuits have also been designed using all-optical switching in the photochromic bacteriorhodopsin protein-coated silica microcavities in contact between two tapered single-mode fibers at telecom wavelengths [18,19]. Although low-power control signals (<200 µW) were used, the speed of operation is ~µs that is governed by the *trans-cis* isomerization of the photosensitive protein. It is important to achieve low-power as well as high switching speed to realize high bit rates, for practical applications.

The objective of this paper is to design basic arithmetic and combinatorial circuits with ultrafast silicon microring resonators, namely (a) Half-Adder/Subtractor and NOR logic gate, (b) Full-Adder/Subtractor, (c) Multiplexer (MUX) and (d) De-Multiplexer (DEMUX) by optimizing the number of switches and inputs, so that they are compact, versatile, lead to simultaneous realization of the desired logic, offer advantages of reconfigurability, lower power consumption and delay. We first present the theoretical model to estimate the intensity based arithmetic and logic operations and then the designs of various computing circuits.

2   Theoretical Model

We consider an all-optical switch in a 2 x 2 silicon add-drop ring resonator as shown in Fig.1.The four ports of the microring resonator switch (M_1) are denoted as Input(I_1), Through(T_1), Add(A_1) and Drop(D_1) according to their functions. When the resonator is on-resonance at λ, a probe signal at IR wavelength λ coupled into ports I_1 and A_1 is directed to D_1 and T_1 respectively. Whereas at off-resonance, light coupled into ports I_1 and A_1 is guided to ports T_1 and D_1, respectively. Considering the

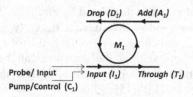

Fig. 1. Configuration of a 2 x 2 all-optical switch based on a silicon add-drop ring resonator

inputs as probe and pump intensities, the all-optical silicon resonator switch can be used as a building block for designing higher computing circuits.

To demonstrate the applicability of the all-optical switching operation and proposed computing circuits, we consider a simplified model neglecting various intrinsic and extrinsic losses. The switching response of the silicon add-drop ring resonator can be modeled considering the propagation of pump and probe beams using coupled-mode theory. A strong optical control pulse and a weak probe light are coupled into the ring resonator through two different resonances. The control pulse generates free carriers in the ring resonator due to two-photon absorption (TPA), which reduce the refractive index of silicon through plasma dispersion effect and blue-shift the ring resonances. The output power of the probe light then gets modulated by the resonance shift. After the control pulse, the resonant wavelength and the transmission of the probe light relax back due to the fast surface recombination of the free carriers [8,9]. The free carrier concentration change is generated predominantly by two-photon absorption and its generation rate is given by [20]

$$\frac{dN(t)}{dt} = \frac{\beta I^2}{2h\nu} - \frac{N(t)}{\tau_{fc}} \tag{1}$$

where, I is the light intensity, $h\nu$ is the photon energy, β is the two-photon absorption coefficient and τ_{fc} is the free carrier recombination lifetime. The induced real refractive index and optical absorption coefficient variations (Δn and $\Delta \alpha$) at a wavelength of 1.55µm produced by free carrier dispersion are given by [20],

$$\Delta n = \Delta n_e + \Delta n_h = -[8.8 \times 10^{-22} \, \Delta N_e + 8.5 \times 10^{-18} (\Delta N_h)^{0.8}] \tag{2a}$$

$$\Delta \alpha = \Delta \alpha_e + \Delta \alpha_h = [8.5 \times 10^{-18} \, \Delta N_e + 6 \times 10^{-18} (\Delta N_h)] \tag{2b}$$

where, Δn_e and Δn_h are the refractive-index changes due to change in electron concentration and hole concentration respectively, $\Delta N_e (cm^{-3})$ is the change in electron concentration, $\Delta N_h \, (cm^{-3})$ is the change in hole concentration, $\Delta \alpha_e (cm^{-1})$ and $\Delta \alpha_h (cm^{-1})$ are the absorption coefficient variations due to ΔN_e and ΔN_h respectively. A change in the carrier density ~ $5 \times 10^{17} cm^{-3}$ can induce a refractive index change (Δn) ~ -1.66 x 10^{-3}, at 1.55µm, causing a shift in the wavelength peak ($\Delta \lambda$) ~1.1 nm [16]. Reflection and Transmission at the output ports are defined as [21],

$$R = \frac{I_{Through}}{I_{Input}} = \left|\frac{E_{reflected}}{E_{input}}\right|^2 = \frac{r_2^2 a^2 - 2r_1 r_2 a \cos \Phi + r_1^2}{1 - 2r_1 r_2 a \cos \Phi + (r_1 r_2 a)^2} \tag{3}$$

$$T = \frac{I_{Drop}}{I_{Input}} = \left|\frac{E_{transmitted}}{E_{input}}\right|^2 = \frac{(1 - r_1^2)(1 - r_2^2)a}{1 - 2r_1 r_2 a \cos \Phi + (r_1 r_2 a)^2} \tag{4}$$

where, r_1 and r_2 are the coupling parameters and a (~1) is the round trip loss attenuation. We consider the incident pump pulse intensity $I = (P/S)\exp[-t^2/\tau^2]$, where P is the peak power, S the effective cross-section and τ is the pulse width. The phase change of the signal is given by $\Delta\Phi = (2\pi/\lambda)\Delta nL$, $(L = 2\pi r)$.

Prior to the arrival of pump pulse, the probe pulse is in high transmission at port T_1 of the silicon microring add-drop resonator. However, the pump pulse changes the resonance wavelength of the resonator, which couples the probe pulse into the resonator, yielding a high transmission at D_1 with output at T_1 getting highly attenuated.

We consider the experimental conditions and results of Almeida *et al.* and Preble *et al.* with the cavity quality factor $Q \cong \lambda_0/\Delta\lambda_{FWHM} = 2290$, at the resonance wavelength $\lambda_0 = 1554.6$ nm and $\Delta\lambda_{FWHM} = 0.68$ nm [16,22]. This quality factor corresponds to a cavity photon lifetime of $\lambda_0^2/2\pi c\,\Delta\lambda_{FWHM} = 1.8$ ps, where c is the speed of light in vacuum. We consider at off-resonance ~98% of light getting transmitted from the I_1/A_1 port to the T_1/D_1 port while at resonance only ~15% getting coupled from the I_1/A_1 port to the T_1/D_1 port [22].

The experimental parameters used for the simulations are : ring radius (r) = 5 μm, probe wavelength (λ_{probe}) = 1554.6 nm, pump wavelength (λ_{pump}) = 400 nm, probe-pulse width = 18 ps, pump-pulse width = 0.1 ps, pump-probe delay = 13.2 ps, photon cavity lifetime = 1.8 ps, effective cross-section of the ring = 450 nm x 250 nm, $r_1 = r_2 = 0.9$, relaxation time (τ_{fc}) = 50 ps, $n_{Si} = 3.48$ [16,22,23].

3   Results and Discussion

The basic switching configuration in Fig.1 has been used to design various computing circuits. The temporal response for the proposed computing circuits has been simulated by considering the experimental conditions and parameters mentioned above.

3.1   Simultaneous Implementation of Half Adder/Subtractor and NOR Logic Gate

A half-adder/subtractor is a circuit that adds and subtracts two bit binary numbers. Fig. 2(a) shows the design of a half-adder/subtractor and a NOR logic gate based on two cascaded add-drop silicon microring resonator switches (M_1 and M_2). The two resonators function as 1 x 2 and 2 x 2 optical switches in the circuit, respectively. Here an optical control signal operating at wavelength 400 nm (which controls the resonant state of the two switches M_1 and M_2) and a probe input at 1554.6 nm are coupled at input port of M_1. The corresponding truth table is shown in Table 1.

When probe Input is switched-off, no light is detected in any of the output ports of the circuit. The control signal is applied to M_1 and M_2 and probe Input is incident at port T_1 of M_1 while $\overline{Input}$ is incident at port A_2 of M_2 (Fig. 2(a)). The first two cases represent the microring resonator in the off-resonant state ($C_1=0$). Case (i): When Input probe is off, T_1 and T_2 are in low state and D_2 is high. Case (ii): When $I_1 = 1$, $T_1 = T_2 = 1$, while $D_1 = D_2 = 0$. Case (iii): When $C_1 = 1$ and $I_1 = 0$, both M_1 and M_2 get enabled (on-resonant). However, since no light is incident at T_1, we have $T_1 = 0$, whereas $T_2 = 1$ and $D_2 = 0$. Case (iv): When both $I_1 = C_1 = 1$, $D_1 = 1$, while $T_2 = D_2 = 0$. This results in the AND operation at D_1, Borrow at T_1, NOR operation at D_2 and Sum/Difference at T_2. We consider the experimental conditions i.e., the pump beam of wavelength 400 nm as the control signal (C_1) and a 200 μW probe laser operating at 1554.6 nm, which butt couples ~ 30% light (60 μW) as the input signal at the Input/Add Port of the silicon microring resonator switch [16,22]. Off-resonance ~98% of light (58 μW) gets transmitted from the Input/Add port to the Through/Drop port while only ~15% (9 μW) is coupled from the Input/Add port to the Through/Drop port on-resonance [22]. A pump beam with a sufficiently narrow bandwidth operating at 5Gits/s bit rate would require an average power of only 11mW [23]. Considering positive logic and an upper limit of low output state to be 15μW and a minimum threshold of 35μW for the high–state for the probe output, Table 1 results in the realization of the Half-Adder/Subtractor and NOR logic operations. Figure 2(b) shows the simulated response of the proposed operation.

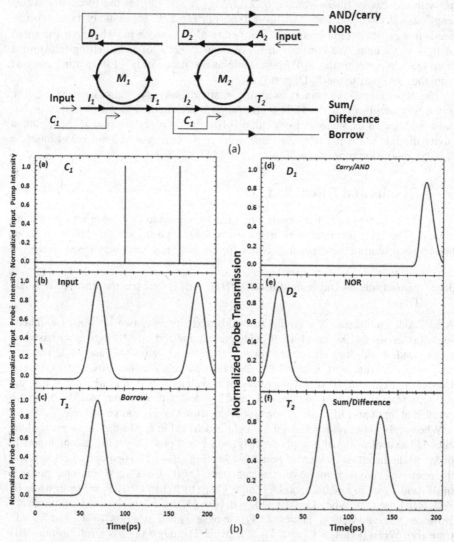

Fig. 2. (a) Schematic of the all-optical Half-Adder, Half-Subtractor and NOR logic circuit, (b) Temporal response: (a) Control Input, (b) Probe Input, (c) Borrow Output, d) Carry/AND Output, (e) NOR Output, and (f) Sum/Difference Output

Table 1. Truth Table of All-Optical Half-Adder/Half Subtractor/NOR Gate

Inputs		Output of Half Adder		Output of Half Subtractor		NOR
Input	Control(C_1)	Sum/XOR	Carry/AND	Difference	Borrow	
0	0	0	0	0	0	1(58µW)
1(60µW)	0	1(58µW)	0	1(58µW)	1(58µW)	0
0	1(11mW)	1(51µW)	0	1(51µW)	0	0
1(60µW)	1(11mW)	0	1(51µW)	0	0	0

3.2 Full-Adder/Subtractor

An all-optical full-adder/subtractor is capable of adding/subtracting three-input bits. To design it, we cascade five microring switches as shown in Fig. 3(a). The probe input and two control inputs are coupled at desired ports of the microring switches.

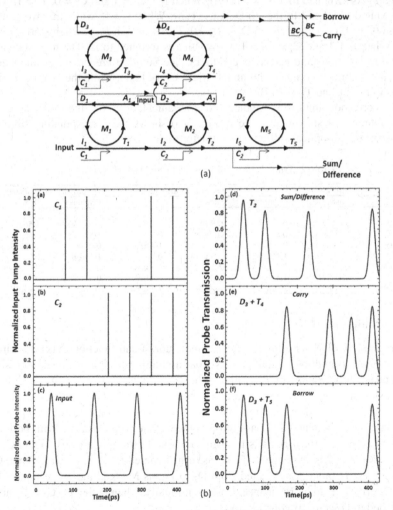

Fig. 3. (a) Schematic of the all-optical Full-Adder and Full-Subtractor logic circuit, (b). Temporal Response: (a) Control Input 1, (b) Control Input 2, (c) Probe Input, (d) Sum/Difference Ouput, (e) Carry Output, and (f) Borrow output. (BC = Beam Combiner)

The first four cases correspond to $C_2 = 0$, while the last four correspond to $C_2 = 1$ that activates M_2, M_4 and M_5. Case (i): When all three inputs are low, we attain a low at all the output ports, since no light is incident on the resonator. When $\overline{\text{Input}}$ is high, it results in $D_1 = T_3 = 1$. Case (ii): When Input probe is on, while the two control beams are off $(C_1 = C_2 = 0)$, light emerges from T_1 to T_2 (Output port for Sum/Difference) resulting in a high output. Case (iii): When $I_1 = C_2 = 0$, $C_1 = 1$ results in $D_1 = D_3 = 0$. Similarly, since $\overline{\text{Input}}$ is given at A_1, it is switched to T_1 and is further directed to T_2 $(C_2 = 0)$. Case (iv): Similarly when probe Input and C_1 are high $(C_2 = 0)$, we obtain $D_1 = 1$, which is redirected to D_3 as well. Case (v): When $I_1 = C_1 = 0$, while $C_2 = 1$, M_2, M_4 and M_5 are on resonance while M_1 and M_3 are deactivated. This results in $T_1 = D_2 = 0$. Since $\overline{\text{Input}}$ is high, a high is attained at D_1, which is directed to A_2 and gets switched to T_2.   Case (vi): When $I_1 = C_2 = 1$ and $C_1 = 0$, $T_1 = 1$, which is transmitted to D_2 and further to D_4. Case (vii): When $I_1 = 0$, while the two controls are high $(C_1 = C_2 = 1)$, we obtain $D_1 = T_2 = 0$. $T_1 = 1$, which gets switched to D_2. Case (viii): When $I_1 = 1$ and $C_1 = C_2 = 1$, $D_1 = 1$ that is rerouted to D_3. The high state at D_1 is also directed to A_2 and gets switched to T_2. The Sum/Difference are obtained at T_2, the beam combined outputs of D_3 and T_4 yield Carry logic, while the beam combined outputs of the D_3 and T_5 yield Borrow logic.

The corresponding truth table for implementation of the full and half adder/subtractor is shown in Table 2. Fig. 3(b) shows the corresponding simulated response of the all-optical full-adder/subtractor logic circuit.

Table 2. Truth Table of All-Optical Full-Adder/Subtractor

Inputs			Output of Full Adder		Output of Half Subtractor	
Input	Control (C_1)	Control (C_2)	Sum	Carry	Difference	Borrow
0	0	0	0	0	0	0
1(60μW)	0	0	1(58μW)	0	1(58μW)	1(56μW)
0	1(11mW)	0	1(50μW)	0	1(58μW)	1(56μW)
1(60μW)	1(11mW)	0	0	1(44μW)	0	1(44μW)
0	0	1(11mW)	1(50μW)	0	1(50μW)	0
1(60μW)	0	1(11mW)	0	1(50μW)	0	0
0	1(11mW)	1(11mW)	0	1(44μW)	0	0
1(60μW)	1(11mW)	1(11mW)	1(44μW)	1(44μW)	1(44μW)	1(44μW)

3.3   Multiplexer (4:1)

A Multiplexer (MUX) is a fundamental combinatorial and function generator circuit, which is used to multiplex input signals from a number of input channels to a single output channel, depending on the value of the select lines. The design of a 4:1 Multiplexer is shown in Fig. 4(a). In this case, the inputs are incident on switches M_1 and M_2 and output is obtained at port D_3. Based on the combination of control inputs, the probe Inputs get switched and pass through the output port. Case (i): With both controls off $(C_1 = C_2 = 0)$, Input 1 is transmitted from T_1 to D_3. Case (ii): When $C_1 = 1$ and $C_2 = 0$, Input 2 is switched to T_1 and is further routed to D_3. Case (iii): When $C_1 = 0$, while $C_2 = 1$, Input 3 is directed from D_2 to D_3, as M_2 is in the off-resonant state, while C_2 drives M_3 to on-resonance. Case (iv): Finally, when both $C_1 = C_2 = 1$, Input 4 is switched to D_2 and subsequently to D_3.

The corresponding truth table for 4:1 MUX is shown in Table 3. Fig. 4(b) shows the corresponding simulated waveform of the all-optical MUX.

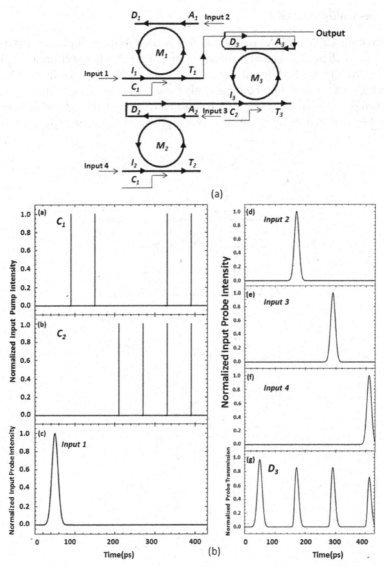

Fig. 4. (a). Schematic of the all-optical 4:1 Multiplexer, (b). Temporal response : (a) Control Input 1, (b) Control Input 2, (c)-(f) Probe Inputs, and (g) Probe Output.

Table 3. Truth Table of All-Optical 4:1 Multiplexer

Data Inputs				Select Inputs/Control Signals		Outgoing Signal
Input 1	Input 2	Input 3	Input 4	Control 1	Control 2	Output
0	X	X	X	0	0	0
1 (60μW)	X	X	X	0	0	1 (58μW)
X	0	X	X	0	1 (11mW)	0
X	1 (60μW)	X	X	0	1 (11mW)	1 (50μW)
X	X	0	X	1 (11mW)	0	0
X	X	1 (60μW)	X	1 (11mW)	0	1 (50μW)
X	X	X	0	1 (11mW)	1 (11mW)	0
X	X	X	1 (60μW)	1 (11mW)	1 (11mW)	1 (44μW)

3.4   De-Multiplexer (1:4)

A De-Multiplexer (DEMUX) is a circuit that is used to de-multiplex an input signal from an input channel to one of the many output channels via select lines. The design of the De-Multiplexer is shown in Figure 5(a). Case (i): When probe Input is low/high and both $C_1 = C_2 = 0$, the probe Input is also low/high as it is transmitted to T_2. Case (ii): When probe Input is low/high, C_1 is low and C_2 is high, the input is directed to T_1 and further switched to D_2. Case (iii): When probe Input is low/high, $C_2 = 0$ while

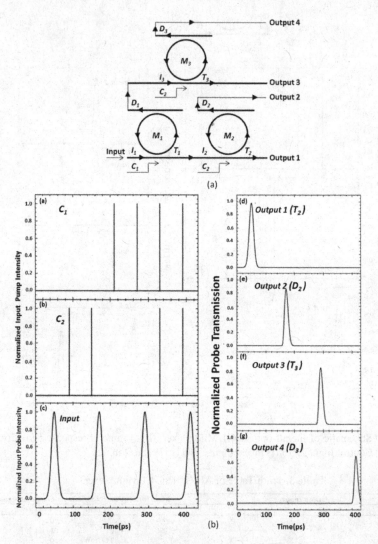

Fig. 5. (a) Schematic of the all-optical 1:4 De-Multiplexer, (b) Temporal response :(a) Control input 1, (b) Control input 2, (c) Probe input, and (d)-(g) Probe Outputs

$C_1 = 1$, probe Input is switched to D_1 and then transmitted to T_3. Case (iv): When I_1 is low/high, $C_1 = C_2 = 1$, probe Input is switched to D_1 and then to D_3. The corresponding truth table for 1:4 DEMUX is shown in Table 4. Fig. 5(b) shows the corresponding simulated waveform of the all-optical DEMUX.

Table 4. Truth Table of All-Optical 1:4 De-Multiplexer

IncomingSignal	Select Inputs/Control Signals		Data Outputs			
Input	Control 1(C_1)	Control2(C_2)	Output 1	Output 2	Output 3	Output 4
0	0	0	0	X	X	X
1(60μW)	0	0	1(58μW)	X	X	X
0	0	1(11mW)	X	0	X	X
1(60μW)	0	1(11mW)	X	1(50μW)	X	X
0	1(11mW)	0	X	X	0	X
1(60μW)	1(11mW)	0	X	X	1(50μW)	X
0	1(11mW)	1(11mW)	X	X	X	0
1(60μW)	1(11mW)	1(11mW)	X	X	X	1(44μW)

Although electro and thermo-optic implementations of various logic circuits with switching speed of ~ns-ms and bit rate 100-10 kbit/sec have been demonstrated, they suffer from lower switching response and bit rates [5,6]. Caulfield and Soref have proposed a reconfigurable electrooptical logic system with SOI resonant structures using dual microring resonators [24]. Although, the circuits facilitate unidirectional signal propagation for cascadability, they require relatively more switches in comparison to the proposed designs, besides adding complexity due to the dual ring structure.

In the proposed designs, the control and probe signals are both considered as logical inputs instead of only control signals in earlier designs [4,7,24]. This reduces the number of switches and inputs to the circuit and results in lower delays and higher bit rates.

MUX /DEMUX operations can be realized from the same circuit by interchanging the inputs and outputs. Advantages of low-power control signals, ultrafast modulation in a silicon resonator, high Q-factor, extinction ratio (~ 12dB), modulation depth (>90%) along with benefits of directed logic, tunability, compactness, high fan-out and large scale integration with flexibility of cascading switches in 2D architectures, make the designs promising for practical applications. The various logic operations require an average optical power ~11 mW for each input control signal and an effective carrier lifetime of 50 ps enables logic operations at ~ 5Gbits/s [23].

Although a simplified model has been used to highlight the applicability of the designs to realize the arithmetic and logic operations, a more rigorous theoretical analysis would be required for practical realization [25]. The proposed logic circuits are general that can be implemented with any externally controlled microring resonator switch in both integrated-optic and fiber-optic formats.

4   Conclusion

We have presented designs of all-optical ultrafast (a) Half-Adder/Subtractor and NOR logic, (b) Full-Adder/Subtractor, (c) Multiplexer and (d) De-Multiplexer circuits, using add-drop silicon microring resonators. The designs require lesser number of switches and inputs for realization of the desired logic compared to other resonator based designs reported earlier in silicon. This leads to lower power consumption, smaller size and latency. The MUX/DEMUX circuits proposed are reconfigurable and scalable. Advantages of directed logic, high Q-factor, tunability, compactness, low-power control signals, high fan-out, and flexibility of cascading switches in 2D architectures to form circuits make the designs promising for practical applications. The designs are general and can be implemented with any other externally controlled microresonator switch.

References

1. Roy, S.: Editorial, Special Issue on Optical Computing Circuits, Devices and Systems. IET Circ., Dev. and Syst. 5, 73–75 (2011)
2. Caulfield, H.J., Dolev, S.: Why future supercomputing requires optics. Nature Photon. 4, 261–263 (2010)
3. Xu, Q., Schmidt, B., Pradhan, S., Lipson, M.: Micrometre-scale silicon electro-optic modulator. Nature 435, 325–327 (2005)
4. Xu, Q., Soref, R.: Reconfigurable optical directed-logic circuits using microresonator-based optical switches. Opt. Exp. 19, 5244–5259 (2011)
5. Zhang, L., Ding, J., Tian, Y., Ji, R., Yang, L., Chen, H., Zhou, P., Lu, Y., Zhu, W., Min, R.: Electro-optic directed logic circuit based on microring resonators for XOR/XNOR operations. Opt. Exp. 20, 11605–11614 (2012)
6. Tian, Y.H., Zhang, L., Ji, R.Q., Yang, L., Zhou, P., Chen, H.T., Ding, J.F., Zhu, W.W., Lu, Y.Y., Jia, L.X., Fang, Q., Yu, M.B.: Proof of concept of directed OR/NOR and AND/NAND logic circuit consisting of two parallel microring resonators. Opt. Lett. 36, 1650–1652 (2011)
7. Lin, S., Ishikawa, Y., Wada, K.: Demonstration of optical computing logics based on binary decision diagram. Opt. Exp. 20, 1378–1384 (2012)
8. Reed, G.T., Mashanovich, G., Gardes, F.Y., Thomson, D.J.: Silicon optical modulators. Nature Photon. 4, 518–526 (2010)
9. Lipson, M.: Guiding, modulating, and emitting light on Silicon-challenges and opportunities. J. Lightwave Technol. 23, 4222–4238 (2005)
10. Bogaerts, W., De Heyn, P., Van Vaerenbergh, T., De Vos, K., Selvaraja, S.K., Claes, T., Dumon, P., Bienstman, P., Van Thourhout, D., Baets, R.: Silicon microring resonators. Laser Photon. Rev. 6, 47–73 (2012)
11. Guha, B., Kyotoku, B.B.C., Lipson, M.: CMOS-compatible athermal silicon microring resonators. Opt. Exp. 18, 3487–3493 (2010)
12. Chin, A., Lee, K.Y., Lin, B.C., Horng, S.: Picosecond photoresponse of carriers in Si ion-implanted Si. Appl. Phys. Lett. 69, 653–655 (1996)
13. Weiss, S.M., Molinari, M., Fauchet, P.M.: Temperature stability for silicon-based photonic bandgap structures. Appl. Phys. Lett. 83, 1980–1982 (2003)
14. Ibrahim, T.A., Cao, W., Kim, Y., Li, J., Goldhar, J., Ho, P.T., Lee, C.H.: All-optical switching in a laterally coupled microring resonator by carrier injection. IEEE Photon. Technol. Lett. 15, 36–38 (2003)
15. Almeida, V.R., Barrios, C.A., Panepucci, R.R., Lipson, M.: All-optical control of light on a silicon chip. Nature 431, 1081–1084 (2004)

16. Almeida, V.R., Barrios, C.A., Panepucci, R.R., Lipson, M., Foster, M.A., Quzounov, D.G., Gaeta, A.L.: All-optical switching on a silicon chip. Opt. Lett. 29, 2867–2869 (2004)
17. Xu, Q., Lipson, M.: All-optical logic based on silicon micro-ring resonators. Opt. Exp. 15, 924–929 (2007)
18. Roy, S., Sethi, P., Topolancik, J., Vollmer, F.: All-optical reversible logic gates with optically controlled bacteriorhodopsin protein-coated microresonators. Adv. Opt. Technol. 2012, 727206-12 (2012)
19. Roy, S., Prasad, M., Topolancik, T., Vollmer, F.: All-optical switching with bacteriorhodopsin protein coated microcavities and its application to low power computing circuits. J. Appl. Phys 107, 053115-24 (2010)
20. Li, C., Na, D.: Optical switching in silicon nano waveguide ring resonators based on Kerr effect and TPA effect. Chin. Phys. Lett. 20, 0542031-4 (2009)
21. Yariv, A.: Universal relations for coupling of optical power between microresonators and dielectric waveguides. Electron. Lett. 36, 321 (2000)
22. Preble, S.F., Xu, Q., Lipson, M.: Changing the color of light in a silicon resonator. Nature Photon. 1, 293–296 (2007)
23. Preble, S.F., Xu, Q., Schmidt, B.S., Lipson, M.: Ultrafast all-optical modulation on a silicon chip. Opt. Lett. 30, 2891–2893 (2005)
24. Caulfied, H.J., Soref, R.A.: Universal reconfigurable optical logic with silicon-on-insulator resonant structures. Photonics Nanostruct. Fundam. Appl. 5, 14–20 (2007)
25. Manolatou, C., Lipson, M.: All-optical silicon modulators based on carrier injection by two-photon absorption. J. Lightwave Technol. 24, 1433–1439 (2006)

Cloud Computing for Nanophotonic Simulations

Nikolay L. Kazanskiy[1] and Pavel G. Serafimovich[2]

[1] Image Processing Systems Institute
of the Russian Academy of Sciences
Samara, Russia
[2] S.P. Korolyov Samara State Aerospace University
Samara, Russia
paulserch@gmail.com

Abstract. Design and analysis of complex nanophotonic and nanoelectronic structures require significant computing resources. Cloud computing infrastructure allows distributed parallel applications to achieve greater scalability and fault tolerance. The problems of effective use of high-performance computing systems for modeling and simulation of subwavelength diffraction gratings are considered. Rigorous Coupled-Wave Analysis (RCWA) is adapted to cloud computing environment. In order to accomplish this, data flow of the RCWA is analyzed and CPU-intensive operations are converted to data-intensive operations. The generated data sets are structured in accordance with the requirements of MapReduce technology.

Keywords: cloud computing, subwavelength diffraction grating, optimization, scatterometry, MapReduce.

1 Introduction

In recent years there has been a steady increase in the research and development activity in the nanoscale fabrication area. Optical methods are widely used to characterize nanoelectronic and optimize nanophotonic structures [1-4]. When the characteristic size of the nanostructure is comparable to the wavelength of the used light, the conventional scalar diffraction methods become inadequate. Therefore, rigorous Maxwell's electromagnetic theory should be adopted to analyze such structures. With the advances of computer technology, many numerical techniques have been developed to rigorously solve the diffraction problem.

Rigorous Coupled Wave Analysis (RCWA) [5-10] has been frequently used to solve the problem of electromagnetic wave diffraction by periodic structures. Solving optimization and scatterometry problems may require millions of diffraction pattern evaluations for each possible combination of geometric structure and incident light parameters. Computational cost estimate shows that these problems would clearly call for supercomputing capacity.

Thus, current computational tasks of nanostructure analysis generate ever-higher demands on the methods used for parallel computing and data storage [11, 12]. The

S. Dolev and M. Oltean (Eds.): OSC 2012, LNCS 7715, pp. 54–67, 2013.

developed applications should work efficiently on multicore and multiprocessor systems. An implementation of MPI (Message Passing Interface) is the traditional means of creating such applications [13]. MPI provides a highly flexible ability to develop applications that address specific requirements of the algorithm structure and the used computing infrastructure. However, this ability makes it necessary for developers to implement some of the low-level services.

Compared with MPI, MapReduce programming paradigm for cloud computing [14] provides higher level of abstraction for the developer of parallel applications. On the one hand, MapReduce imposes certain restrictions on data format and the flexibility of the algorithm used. On the other hand, MapReduce offers a simple programming model, automatic parallelization and distribution, fault-tolerance, I/O scheduling, monitoring tools.

In the following we consider how RCWA can be implemented to utilize MapReduce technology. The RCWA relies heavily on the computation of eigenvalues of an intermediate matrix and the solution of a corresponding linear system. To reduce the compute time and enhance fault tolerance, we build a distributed, dynamically growing look-up table containing the precomputed eigenvalues and eigenvectors for the set of lamellar layers that approximate the analyzed nanostructure. The look-up table resides in a specialized distributed file system.

The paper is organized as follows. Section II describes the main features of MapReduce framework implementation for cloud computing. Section III outlines the algorithm and data flow of RCWA. Section IV reports the adaptation of RCWA for MapReduce technology. Section V describes some performance tests for the suggested algorithm on the Hadoop-cluster [14, 16]. Conclusions are given in the final section.

2   Main Features of MapReduce Technology

MapReduce is based on the old approach of splitting a task into subtasks, executing them in parallel and merging the intermediate results to obtain the final one [17-19]. However, the MapReduce implementations of this scheme provide effective means for controlling resource utilization in large scale distributed systems.

The processing of large data sets (more than hundreds of gigabytes) is a challenge for the traditional High Performance Computing (HPC) attitude. In comparison to CPU-intensive tasks, the data-intensive tasks require demanding rate of access to the used storage system. The bottleneck of HPC for these problems is the shared files system network bandwidth. Hence, a large number of computing nodes would remain idle and wait for the data. Because of this, the load balancing and data locality become important to efficiently handle.

MapReduce framework provides load balancing of computing nodes by organizing the "master-workers" architecture. The master node is responsible for the assignment of computing tasks to worker nodes every time a worker node becomes idle.

For a large data set it is crucial to ensure the closeness of data and computation to reduce data transfer between nodes. Thus, the input data should already be stored locally on the corresponding computing nodes. MapReduce achieves this by using a specialized distributed file system. This file system divides each file into blocks with size of about several tenths of megabytes and stores several copies of each block on different nodes. This data replication in the most cases allows the master node to assign computing tasks to a worker that stores corresponding input data.

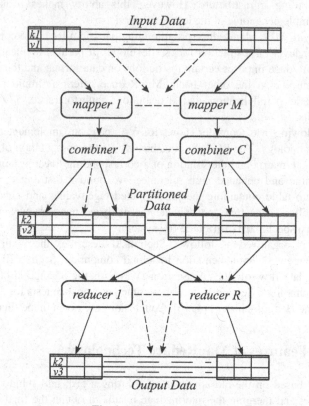

Fig. 1. MapReduce scheme

The next problem of traditional HPC is the difficulty in handling partial failure of computing nodes. In large clusters of thousands of computing nodes it is necessary to permanently deal with crashes. In MapReduce, the master node pings every worker periodically to recognize possible failures of individual nodes. Thus, master detects a failed computing task and reschedules it to another worker which owns a replica of the input data block.

The two-stage processing scheme of MapReduce is depicted in Fig. 1. The first and second stages correspond, respectively, to Map() and Reduce() functions. Both functions need to be implemented by the developer of distributed application. Key/value pairs form the required processing data structure. Every input key/value pair is processed by the Map() to produce a set of intermediate key/value pairs. The Reduce() takes all values associated with the same intermediate key as input and generates a set of final key/value pairs. Thus, it is sufficient to implement the two functions to obtain a usable distributed application.

To reduce the network traffic between nodes MapReduce supports the use of auxiliary Combine() function which is similar to Reduce(). Combine() processes the output of the Map() on each node of the cluster separately to prepare the intermediate results to subsequent shuffling over a network to the Reduce() nodes.

The shuffling of the intermediate results is based on the dividing up the set of intermediate keys and assigning intermediate key/value pairs to Reduce() nodes. The goal is to assign approximately the same number of keys to each Reduce() node.

3   Overview of RCWA

The general grating diffraction problem is illustrated in Fig. 2. It is convenient to separate the space into three regions: two homogeneous semi-infinite regions above and below the grating, and an inhomogeneous region which includes the relative permittivity periodic modulation of the analyzed nanostructure. A linearly polarized plane electromagnetic wave of wavelength λ is obliquely incident at an arbitrary polar angle θ and at an azimuthal angle ϕ upon a two-dimensional, possibly multi-level or surface-relief, dielectric or metal grating. The normalized electric-field vector that corresponds to this plane wave is a solution of Maxwell's equations in an infinite homogeneous region [20]. The variable ψ is the angle between the polarization vector E and the plane of incidence.

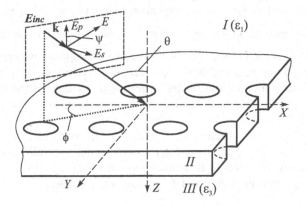

Fig. 2. Analyzed nanostructure

RCWA was formulated for both principal plane and conical incidence [6]. The problem of the slow convergence of the solution when a conducting grating is illuminated with a transverse magnetic (TM) polarized wave has been subsequently resolved [7].

RCWA involves several steps. As shown in Figure 3(a), the grating is sliced into a set of two-dimensional layers and each layer is approximated by a rectangular slab. The periodic permittivity distribution in each slab of the grating region is expanded into the Fourier series.

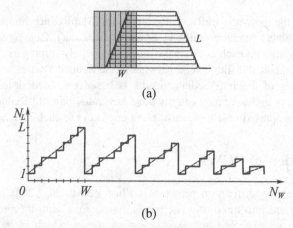

Fig. 3. Example of two-dimensional diffraction grating analysis; (a) the grating period with the admissible value ranges of the upper and lower radius of the truncated cone; (b) the number of eigenvalues and eigenvectors calculations for each set of the truncated cone upper and lower radius without storing the intermediate information

The field in the grating region can be also expressed as Fourier expansion through application of the Floquet condition. Substituting these expansions into the Maxwell equations, we can derive a set of first-order coupled wave equations. Its homogeneous solution can be obtained by solving for the eigenvalue and eigenvector of the matrix that corresponds to this set of wave equations. The size of the matrix is equal to $(2N+1)^4$ for conical incidence, where N is the number of positive observed diffraction orders for each of the coordinates X and Y. The electromagnetic field above and below the grating region can be expressed using linear superposition of plane waves in the direction of various orders of grating diffraction. In the final step, this representation is used as the boundary conditions to determine the specific solution of the diffraction efficiency.

Accuracy of RCWA is related to the number of diffraction orders that are included in the superposition. In general, the greater number of observed diffraction orders the higher the accuracy. However, it is necessary to control the systematic computational error when a large number of observed diffraction orders is selected.

4   Mapping RCWA to the MapReduce Scheme

We consider the technique of using the RCWA for optical simulation or optimization of the nanostructures. As an example of such structure, we take a two-dimensional diffraction grating. The grating period consists of a truncated cone, and its central cross section is shown in Fig. 3 (a). The gray rectangle in Fig.3 (a) shows the set of the possible (evaluated) geometric parameters. The slope of the truncated cone lies inside this rectangle. The parameter W determines the number of vertically homogenous layers that approximate the slope in RCWA. We have the number of these layers is fixed and equal to L.

As mentioned in the preceding section, eigenvalues and eigenvectors of the corresponding matrix are calculated for each vertically homogeneous layer. The calculated eigenvalues and eigenvectors do not depend on the thickness of a homogeneous layer. Without saving the intermediate results, this time-consuming task would be repeated many times. The sawtooth step function in Fig. 3 (b) shows the dependence of number N_L of eigenvalues and eigenvectors calculations without storing the intermediate information on N_W sets of radius variation. The parameter L in the Fig. 3 (b) is assumed to be equal to 7 and parameter W is equal to 8. Note that the number of eigenvalues and eigenvectors calculations is equal to W when the intermediate information is stored.

Such storage of the intermediate information converts the CPU-intensive problem into a data-intensive one. We estimate the amount of memory needed to store the calculated eigenproblem data.

The structure and number of input parameters for the task of optical modeling of nanostructures depends in particular on the simulation purposes and the geometry of the structure. Thus, we can distinguish two main groups of input parameters. The first group determines the illumination conditions of the simulation. The second group consists of the geometric parameters of the nanostructure and its variations.

We estimate the possible number of input parameters for the task of modeling the nanostructure as shown in Figure 3 (a). We start with the group of geometrical parameters. Let the number of possible values of the truncated cone radius be 100 ($W =$ 100). Suppose that a variation of the structure period is also allowed. The number of possible values of the structure period is equal to 10. Let us now consider the group of the illumination conditions parameters. Suppose that in our case this group contains two parameters. First, it is the meridional angle of incidence. Second, it is the azimuth angle of incidence. For each of these two parameters, we set the number of possible variations to 30. Thus, the number of eigenproblems to solve is equal to the product of the number of variations of the aforesaid four parameters, i.e. $\sim 10^6$.

Having computed the parameters number, we now evaluate the size of the stored intermediate data. This value depends on the number of observed diffraction orders. Next we assume that the calculations are done with long double precision (16 bytes).

Fig. 4 (a) shows the dependence of the size of the eigenvectors matrix on the number of observed diffraction orders. Fig. 4 (b) shows the total amount of the stored intermediate data for the different number of the observed diffraction orders and the parameters numbers. The figure shows that the amount of data may exceed 1 TB. Thus, our task can be classified as a data-intensive problem. To reduce the size of the stored intermediate data some compression procedure can be used. However, this in turn implies there is a corresponding computational overhead.

We will now consider how RCWA can be mapped into MapReduce scheme. The structure of stored data for MapReduce technology is defined by a pair of "key /

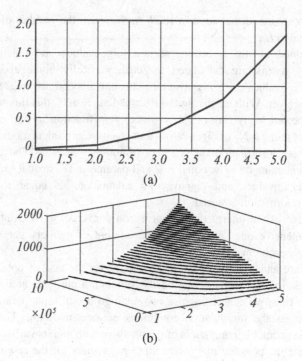

Fig. 4. The size of the stored intermediate data; (a) the dependence of the size of the eigenvectors matrix on the number of observed diffraction orders; (b) the total amount of the stored intermediate data for the different number of the observed diffraction orders and the parameters numbers.

value". Here the "key" $\mathbf{K}_1$ consists of two sets. First, it is the set of geometric parameters that define one of the layers which is homogeneous in the vertical direction. Second, it is the set of parameters that define the illumination conditions of the simulation. The following relation is an example of such "key".

$$\mathbf{K}_1 = \left\{ d_x, m_x, \mathbf{b}_x, \mathbf{n}_x, d_y, m_y, \mathbf{b}_y, \mathbf{n}_y, \lambda, \theta, \phi \right\}$$

where d_x, d_y are the periods of two-dimensional structure, m_x, m_y are the number of material sections on the period, $\mathbf{b}_x, \mathbf{b}_y$ are the size of material sections on the period, $\mathbf{n}_x, \mathbf{n}_y$ are the index of refraction of these material sections. The last three parameters in the "key" - λ, θ, ϕ - describe the lighting conditions. Notice that the thickness of the layer is absent in the above list.

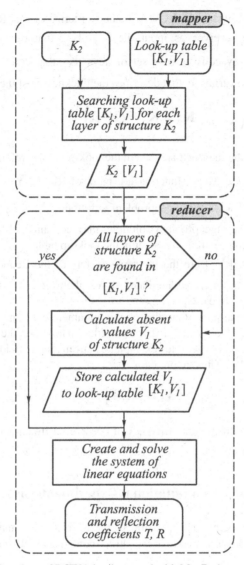

Fig. 5. Flowchart of RCWA implemented with MapReduce technology

The "value" V_1 contains the calculated eigenvalues and eigenvectors for one layer:

$$V_1 = V_2 = \{Eig\}$$

The look-up table $K_1 / [\, V_1 \,]$ resides in the distributed file system mentioned in Section II. By combining with MapReduce, the distributed file system allows providing efficient load balancing, data locality and faulting tolerance to the whole computing system.

Fig. 5 shows the flowchart of RCWA method in terms of MapReduce technology. The Map() function receives as input the key $\mathbf{K}_2$ that describes the analyzed nanostructure. This key consists of a set of keys $\mathbf{K}_1$ that define the layers of the structure, as well as the array $\mathbf{t}$ that describes the thicknesses of these layers.

$$\mathbf{K}_2 = \mathbf{K}_3 = \left\{ \ddot{\mathbf{K}}_1, \mathbf{t} \right\}$$

The Map() looks for the intermediate data in the look-up table with the given key $\mathbf{K}_2$. Thus, the output of the Map() function are pairs of $\mathbf{K}_2 / [\mathbf{V}_2]$. Each structure is associated with the list of the eigenproblem solutions $[\mathbf{V}_2]$ for the array of $\ddot{\mathbf{K}}_1$ layers. This list of the eigenproblem solutions may contain null values, in the case where the necessary information is not yet in the look-up table.

The pairs of $\mathbf{K}_2 / [\mathbf{V}_2]$ are the input parameters of the Reduce(). This function calculates the missing intermediate data, stores them in the look-up table, then generates the system of linear equations and solves it.

In this paper, we assume that the output parameters of the Reduce() are the diffraction efficiency of observed diffraction orders. However, the same approach is valid for other output parameters, for example, electromagnetic field distribution in the so-called near-field [10]. Thus, in our case

$$\mathbf{V}_3 = \left\{ \mathbf{R}, \mathbf{T} \right\}.$$

where $\mathbf{R}, \mathbf{T}$ are the diffraction efficiency of observed diffraction orders respectively for reflection and transmission.

5   The Results of Computational Experiments on Hadoop-Cluster

Computational experiments are performed on a relatively small Hadoop-cluster consisting of four nodes. Each node in the cluster contains a CPU Intel Xeon 2.13 GHz and 4 GB RAM.

Namenode-server is run on one node, and jobtracker-server is run on another node. The remaining two cluster nodes are used as working nodes. Two mappers and two reducers can be run simultaneously on each of the working nodes, which correspond to the parameters that are set by default. Other parameters of Hadoop-cluster are also set by default.

The experiment measures the parallel read/write average throughput for the HDFS file system. The parallel read average throughput is equal to $r=460$ MB/s and the parallel write average throughput is equal to $w=60$ MB/s. It is not our goal in this work

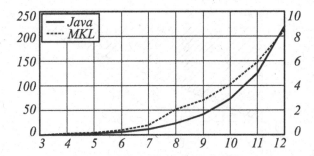

Fig. 6. The dependence of the time (in seconds) of eigenvectors calculation on N value for Java-implementation (the left Y-axis) and Intel MKL implementation (the right Y-axis)

to enhance r and w values by optimizing the default parameters of Hadoop-cluster. Therefore, the above values r and w are assumed to be fixed.

We will now assess the feasibility of the proposed scheme of RCWA implementation. The standard and the proposed algorithms are different only at the eigenvectors calculation step. We compare the time of eigenvectors calculation for the standard RCWA algorithm (without saving the intermediate results) with the time of eigenvectors calculation for the proposed algorithm (with saving the intermediate results). The main share of the RCWA computational time is contributed by the eigenvectors calculation. The eigenvectors are calculated for each layer of the modeled nanostructure.

Java programming language is the most appropriate for the implementation of a flexible application within the Hadoop package. Therefore, the algorithm for finding the eigenvectors of a complex general matrix was implemented with Java.

Figure 6 shows the dependence of the time (in seconds) of eigenvectors calculation on N value for Java implementation and Intel MKL implementation. N value denotes the number of the positive diffraction orders for one coordinate. Here we assume that N value is the same for each coordinate.

The left Y-axis on the graph shown in the Figure 6, corresponds to the results of Java-application. The right Y-axis of this graph shows the same results for Intel MKL implementation.

We define the following parameters:

s - Upload file size (MB), which is processed by one mapper;

f - The "useful" share of the uploaded file, which is used by mapper to find the eigenvectors (normalized value);

m - The average number of uses of one matrix from the uploaded file;

a - The size of the matrix (MB), which contains the computed eigenvectors for one layer of the modeled nanostructure;

e - The time for computing the eigenvectors of the matrix for one layer of the structure (s);

L - The total number of layers in structures that are processed by one mapper.

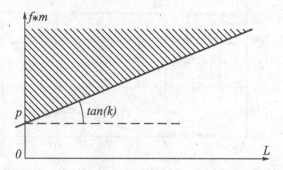

Fig. 7. The dependence of L on the overall usage ratio of the uploaded file; the cross hatched area shows where the relation (1) is satisfied

Then the condition for the feasibility of the proposed RCWA algorithm can be written as follows: the total computational time that one mapper spends *(L e)* should be greater than the time spent on the following steps: (1) the reading from the HDFS file with size of *s*, (2) the calculation of the unresolved eigenvectors, (3) the writing of the calculated eigenvectors in HDFS.

The above described condition can be represented as a linear dependence of the *L* value on the overall usage ratio of the uploaded file *f m*:

$$(f \cdot m) > k \cdot L + p \tag{1}$$

where

$$k(N) = \frac{a^2(N)}{s(e(N) \cdot w - a(N))}, \quad p(N) = \frac{a(N) \cdot w}{r(e(N) \cdot w - a(N))}.$$

In Fig. 7 the cross hatched area shows where the relation (1) is satisfied. The slope of the curve in the Figure 7 (*k* value) defines the rate at which the overall usage ratio *f m* should increase when *L* value increased.

Fig. 8 shows the dependence of the curve slope value *k* on *N* value for (a) Java-implementation and (b) Intel MKL implementation.

Note that the curve slope value *k* is very small on the Figure 8(a). Thus, to stay within the feasibility requirements, there is no much importance struggling for the increasing the overall usage ratio *f m* when using the relatively slow Java-implementation. Such need appears for the more effective Intel MKL implementation (Fig. 8(b)). In this case, the different approaches of meta-data usage can be considered.

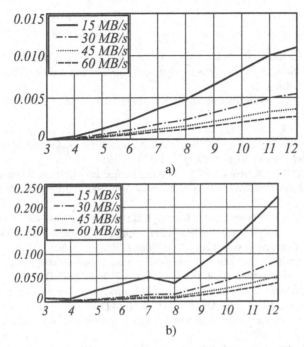

Fig. 8. The dependence of k value on the number of diffraction orders value N for (a) Java-implementation and (b) Intel MKL implementation

6   Conclusions

The paper discusses the computational problems arising in modeling and optimization of complex nanophotonic structures by Fourier modes method (RCWA). The cloud computing infrastructure usage is suggested to solve these problems. This approach allows to effectively exploit the potential of modern computational tools by improving the scalability of the computational problem and by enhancing the fault tolerance of the computer system. This opens up new possibilities in solving problems of diffraction nanophotonic [21-23], magneto-optics [24,25] and plasmonics [26,27].

Acknowledgments. This work is supported by the program of the RAS Presidium "Problems of a national environment for distribution of scientific information and computing based on the development of GRID technology and modern telecommunications networks", grant of Russian Federation President for Support of Leading Scientific Schools NSh-4128.2012.9 and RFBR grants No 11-07-00153, 13-07-97002, 13-07-97004, program No 5 of basic research for Nanotechnology and Information Technology Department of RAS "Basic problems of physics and technology of epitaxial nanostructures and related devices".

References

1. Golub, M.A., Kazanskii, N.L., Sisakyan, I.N., Soifer, V.A.: Computational experiment with plane optical elements. Optoelectronics, Instrumentation and Data Processing (1), 78–89 (1988) (in Russian)
2. Kazanskiy, N.L.: Mathematical simulation of optical systems. SSAU, Samara (2005) (in Russian)
3. Kazanskiy, N.L., Serafimovich, P.G., Khonina, S.N.: Harnessing the guided-mode resonance to design nanooptical transmission spectral filters. Optical Memory & Neural Networks (Information Optics) 19(4), 318–324 (2010)
4. Golovashkin, D.L., Kazanskiy, N.L.: Solving Diffractive Optics Problem using Graphics Processing Units. Optical Memory and Neural Networks (Information Optics) 20(2), 85–89 (2011)
5. Moharam, M.G., Pommet, D.A., Grann, E.B.: Stable implementation of the rigorous coupled-wave analysis for surface-relief gratings: Enhanced transmittance matrix approach. J. Opt. Soc. Am. A 12(5), 1077–1086 (1995)
6. Gystis, E., Gaylord, T.: Three-dimensional (vector) rigorous coupled wave analysis of anisotropic grating diffraction. J. Opt. Soc. Am. A 7, 1399–1419 (1990)
7. Lalanne, P., Morris, G.M.: Highly improved convergence of the coupled-wave method for TM polarization. J. Opt. Soc. Am. A 13(4), 779–784 (1996)
8. Li, L.: Use of Fourier series in the analysis of discontinuous periodic structures. J. Opt. Soc. Am. A 13(9), 1870–1876 (1996)
9. Bezus, E.A., Doskolovich, L.L., Kazanskiy, N.L.: Evanescent-wave interferometric nanoscale photolithography using guided-mode resonant gratings. Microelectronic Engineering 88(2), 170–174 (2011)
10. Bezus, E.A., Doskolovich, L.L., Kazanskiy, N.L.: Scattering suppression in plasmonic optics using a simple two-layer dielectric structure. Applied Physics Letters 98(22), 221108 (3 p.) (2011)
11. Armbrust, M., et al.: A view of cloud computing. Communications of the ACM 53(4), 50–58 (2010)
12. Volotovskiy, S.G., Kazanskiy, N.L., Seraphimovich, P.G., Kharitonov, S.I.: Distributed software for parallel calculation of diffractive optical elements on web-server and cluster. In: Proc. IASTED, pp. 69–73. ACTA Press (2002)
13. Snir, M., Otto, S., Huss-Lederman, S., Walker, D., Dongarra, J.: MPI-The Complete Reference. The MPI Core, vol. 1. MIT Press, Cambridge (1998)
14. Dean, J., Ghemawat, S.: MapReduce: a flexible data processing tool. Communications of the ACM 53(1), 72–77 (2010)
15. http://Hadoop.apache.org/ (Tested June 15, 2011)
16. Venner, J.: Pro Hadoop. Springer (2009)
17. Voevodin, V.V.: Mapping computational problems in computer architecture. Computational Methods and Programming: New Information Technologies 1(2), 37–44 (2000) (in Russian)
18. Popov, S.B.: Modeling the task information structure in parallel image processing. Computer Optics 34(2), 231–242 (2010) (in Russian)
19. Soifer, V.A. (ed.): Computer Image Processing, Part I: Basic concepts and theory, 283 p. VDM Verlag Dr. Muller e.K. (2009)
20. Born, M., Wolf, E.: Principles of Optics. Pergamon, Oxford (1980)
21. Soifer, V.A.: Nanophotonics and diffractive optics. Computer Optics 32(2), 110–118 (2008) (in Russian)

22. Soifer, V.A., Kotlyar, V.V., Doskolovich, L.L.: Diffractive optical elements in nanophotonic devices. Computer Optics 33(4), 352–368 (2009) (in Russian)
23. Kazanskiy, N.L., Serafimovich, P.G., Popov, S.B., Khonina, S.N.: Using guided-mode resonance to design nano-optical spectral transmission filters. Computer Optics 34(2), 162–168 (2010) (in Russian)
24. Belotelov, V.I., Doskolovich, L.L., Zvezdin, A.K.: Extraordinary magneto-optical effects and transmission through metal-dielectric plasmonic systems. Physical Review Letters 98(7), 5 p. (2007)
25. Bykov, D.A., Doskolovich, L.L., Soifer, V.A., Kazanskiy, N.L.: Extraordinary Magneto-Optical Effect of a Change in the Phase of Diffraction Orders in Dielectric Diffraction Gratings. Journal of Experimental and Theoretical Physics 111(6), 967–974 (2010) (in Russian)
26. Bezus, E.A., Doskolovich, L.L., Kazanskiy, N.L., Soifer, V.A., Kharitonov, S.I., Pizzi, M., Perlo, P.: The design of the diffractive optical elements to focus surface plasmons. Computer Optics 33(2), 185–192 (2009) (in Russian)
27. Bezus, E.A., Doskolovich, L.L., Kazanskiy, N.L., Soifer, V.A., Kharitonov, S.I.: Design of diffractive lenses for focusing surface plasmons. Journal of Optics 12(1), 015001 (7 p.) (2010)

All-Optical Ultrafast Switching and Logic
with Bacteriorhodopsin Protein

Sukhdev Roy[*] and Chandresh Yadav

Department of Physics and Computer Science
Dayalbagh Educational Institute, Dayalbagh, Agra 282 110 India
sukhdevroy@dei.ac.in

Abstract. We present a detailed analysis of all-optical ultrafast switching with the unique photochromic bacteriorhodopsin (bR) protein, based on its early transitions ($B_{570} \rightarrow I_{460}$), in the pump-probe configuration. The transmission of a cw probe laser beam at 460 nm through bR is switched by a pulsed pump beam at 570 nm with high contrast and sub-ps switching. The effect of pump intensity, pump pulse width, absorption cross-section and lifetime of the I_{460} state on the switching characteristics has been studied in detail. Theoretical simulations are in good agreement with reported experimental results. The results have been used to design ultrafast all-optical NOT and the universal NOR and NAND logic gates with multiple pump laser pulses. The analysis demonstrates the applicability of bR for all-optical ultrafast operations in the simple pump-probe geometry and opens up exciting prospects for its use in optical supercomputing.

Keywords: all-optical switching, ultrafast information processing, optical computing, logic gates, bacteriorhodopsin.

1 Introduction

The anticipated need for ultrafast and ultrahigh bandwidth information processing has provided impetus to realize all-optical information processing [1,2]. All-optical devices require control of light with light in a material that exhibits an efficient nonlinear optical response. A wide variety of organic molecules have attracted considerable attention due to their ultrafast optical response, high nonlinearities and the ability to tailor their properties to meet practical requirements [3-6]. However, since light sustains life on earth, a wide range of biomolecules have evolved with an optimized energy efficient photoresponse to control various kinds of photo-induced molecular responses that include the life sustaining process of photosynthesis.

The photochromic protein bR found in the purple membrane fragments of *Halobacterium salinarum* is a model system to understand the signal transduction mechanism in nature and has emerged as an outstanding material for biomolecular photonic applications due to its unique properties [7]. It exhibits a multifunctional photoresponse, namely, photochromism, photoelectricity and proton pumping. It exhibits high quantum efficiency of converting light into a state change (~64%),

---

[*] Corresponding author.

S. Dolev and M. Oltean (Eds.): OSC 2012, LNCS 7715, pp. 68–77, 2013.

large absorption cross-section and nonlinearities, robustness to degeneration by environmental perturbations, capability to form thin films in polymers and gels, and flexibility to tune its kinetic and spectral properties by genetic engineering techniques, for device applications [7]. It also exhibits high photo and thermal stability over a wide pH range (0-12). bR molecules in gels and polymer thin films are extremely stable, maintaining their functionality for several years without any degradation. No noticeable change has been observed in a bR film even after switching more than a million times [7]. In addition, dry bR films are structurally stable upto a temperature of 140 ^{0}C [8].

By absorbing green–yellow light, the wild-type bR molecule undergoes several conformational transformations in a complex photocycle that generates a number of intermediate states, spanning the entire visible spectrum. The main photocycle exhibits the following transformations: B_{570} + hv → J_{625} (< ps) → K_{610} (~µs) → L_{540} (~µs) → M_{410} (~ms) → N_{550} (~ms) → O_{640} (~ms) → B_{570}, where subscripts denote the respective peak absorption wavelengths in nm (Fig.1).

A wide range of applications has been proposed with bR that include energy conversion, artificial retina and 3D memories. Various designs of all-optical switching and logic gates have been proposed with bR based on its photochromic properties that include sequential photoexcitation [9], two wave mixing [11], degenerate four wave mixing (DWFM) [12], photoinduced dichroism and birefringence [13], complementary modulated suppression transmission (CSMT) [14], degenerate multiwave mixing (DMWM) [15] and nonlinear intensity-induced excited-state absorption [16-18]. Most of the photonic applications of bR have been proposed by utilizing its longer lifetime M and O intermediates resulting in ms switching [10,16-18]. Recently, all-optical switching and arithmetic logic operations have been shown with optically controlled bR-coated microresonators with µs switching response [19,20]. Der *et al.* and Fabian *et al.* have demonstrated switching in the ns range based on B_{570} → K_{610} transition [21,22]. The main limitation with optical processing with bR has been the speed, although the operations are at much lower powers compared to other materials.

Recently, femtosecond spectroscopic characterization of the initial *trans-cis* isomerization in the bR photocycle has revealed additional spectrally distinct states before the J_{625} state: B_{570} + hv → H_{FC} (~fs) → I_{460} (FS) (<ps) → J_{625}. Although, detailed experiments have been performed to ascertain the signal transduction pathways and characterization of these states, a complete picture is still under development. H_{FC} denotes the Franck-Condon excited state of bR with an *all-trans* retinal Schiff base and I_{460} denotes the fluorescent state (FS), which is characterized by absorption around 460 nm and a stimulated emission band around 860 nm. H_{FC} and I_{460} are excited-states with a chromophore of an *all-trans* conformation, whereas, the K_{610} state has the chromophore converted to the *13-cis* conformation. It has been ascertained that the H_{FC} state exhibits a fluorescence Stokes shift as it relaxes in ~100 fs to the I_{460} excited-electronic state of the protonated Schiff base chromophore, which subsequently can relax to the J_{625} and then to K_{610} in 500 fs and 3ps, respectively or return directly to the *all-trans* ground state [23-33]. Attempts have been made to explain the primary molecular process by a two state model. However, experiments and *ab initio* calculations have favoured a three state model with changes in the molecules structure assumed to precede the *trans-cis* isomerization [28]. These early sub-ps transitions open up possibilities of ultrafast all-optical switching and computing with bR. Fabian *et al.* have recently demonstrated ultrafast

switching based on shift in the frequency of a probe laser beam in an integrated-optic planar geometry, with a bR adlayer deposited above a 1D photonic crystal, on a thin film waveguide carried by a glass substrate [34].

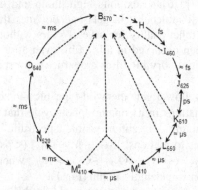

Fig. 1. Photocycle of bacteriorhodopsin. Solid and dashed arrows represent thermal and photoinduced transitions, respectively.

In this paper, we theoretically analyze all-optical ultrafast switching based on the $B_{570} \rightarrow I_{460}$ transitions in the versatile pump-probe configuration and investigate its application to computing. The effect of various parameters such as pump intensity, pump pulse width, pulse frequency, absorption cross-section and life time of the I_{460} state on the switching characteristics has been studied in detail. Further, switching has been optimized to design all-optical ultrafast NOT and the universal NOR and NAND logic gates that are the basic building blocks of computing circuits.

2   Theoretical Model

A schematic diagram for pump-probe configuration is as shown in Fig. 2(a), in which the intensity of a modulating pump pulse switches the transmission of a weak cw probe beam in a bR sample. We introduce a simplified level diagram shown in Fig. 2(b) to represent the ultrafast early transitions in the photochemical cycle of bR molecules, which enables adoption of the simple rate-equation approach for the population densities in the various intermediate states. We consider bR molecules exposed to a pump beam of intensity I_m, which modulates the population densities of different states through the excitation and de-excitation processes that can be described by the rate equations in the following form,

$$\frac{dN_B}{dt} = -\frac{\sigma_B I_m N_B \psi_{BH}}{h\nu} + \frac{\sigma_I I_m N_I}{h\nu} + \frac{N_H}{\tau_0} + \frac{N_I}{\tau_2} \qquad (1)$$

$$\frac{dN_H}{dt} = \frac{\sigma_B I_m N_B \psi_{BH}}{h\nu} - \frac{N_H}{\tau_0} - \frac{N_H}{\tau_1} \qquad (2)$$

$$\frac{dN_I}{dt} = -\frac{\sigma_I I_m N_I}{h\nu} + \frac{N_H}{\tau_1} - \frac{N_I}{\tau_2} \qquad (3)$$

where N_B, N_H and N_I are the population densities of the B, H and I states, respectively, σ_B and σ_I are the absorption cross-sections of the B and I states at modulating pump wavelength, respectively. ψ_{BH} is the quantum efficiency for B→H transition and τ_0, τ_1 and τ_2 are the relaxation times for non-radiative transitions.

Assuming optically thin bR samples, the propagation effects on the modulating light beams can be neglected. The modulating pump laser pulse is given by,

$$I_m = I_{m0} \exp\left(-c\left(\frac{t-t_m}{\Delta t}\right)^2\right) \tag{4}$$

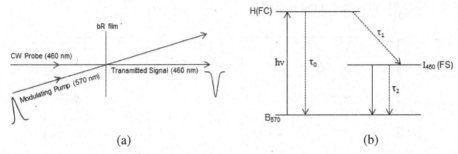

(a)   (b)

Fig. 2. (a). Schematic diagram for all-optical switching, (b) Simplified energy level diagram of $B_{570} \rightarrow I_{460}$ transitions in bR

To analyze two input logic gates configurations, we consider two pump pulses such that the four input combinations (0,0), (0,1), (1,0) and (1,1) in terms of phase differences of their peaks are given by [10],

$$I_{m1} = I_{m0}\left[\exp\left(-c\left(\frac{t-t_{m1}}{\Delta t}\right)^2\right) + \exp\left(-c\left(\frac{t-t_{m2}}{\Delta t}\right)^2\right)\right] \tag{5}$$

and

$$I_{m2} = I_{m0}\left[\exp\left(-c\left(\frac{t-t_{m3}}{\Delta t}\right)^2\right) + \exp\left(-c\left(\frac{t-t_{m2}}{\Delta t}\right)^2\right)\right] \tag{6}$$

where I_{m0} is the peak pumping intensity, t_m, t_{m1}, t_{m2} and t_{m3} are the times at which the respective pulse maxima occur, c is the pulse profile parameter and Δt is the pulse width. We consider the transmission of a cw probe laser beam of intensity $I_p(I_p<<I_m)$ at 460 nm corresponding to the peak absorption of the I_{460} intermediate in the bR photocycle, modulated by absorption due to excitation of bR molecules by the pulsed pump laser beam at 570 nm. The nonlinear intensity dependent absorption coefficient for the probe beam is written as $\alpha_p(I_m) = N_I(I_m)\sigma_{Ip}$, where x is the distance in the medium and σ_{Ip} is the absorption cross-section at probe wavelength. The propagation of the cw probe laser beam through the bR medium is governed by,

$$\frac{dI_p}{dx} = -\alpha_p(I_m)I_p \tag{7}$$

3   Results and Discussion

The optical switching characteristics, namely the variation in the normalized transmitted intensity of the probe laser beam with time have been computed by solving the rate equations for the early intermediate states through computer simulations, using equations (1)-(7), considering the spectroscopic data and experimental conditions, with $\sigma_B = 2.4 \times 10^{-16}$ cm^2, $\sigma_I = 0.22 \times 10^{-16}$ cm^2, $\sigma_{Ip} = 1.8 \times 10^{-16}$ cm^2 at probe wavelength, $\tau_0 = 30$ fs, $\tau_1 = 100$ fs, $\tau_2 = 315$ fs, $\Delta t = 120$ fs, film thickness $L = 30$ µm [10,23-30]. The variation in the normalized transmitted probe at 460 nm corresponding to the peak absorption of the I_{460} intermediate with time for different peak pump intensity (I_{m0}) values at 570 nm is shown in Fig. 3(a).

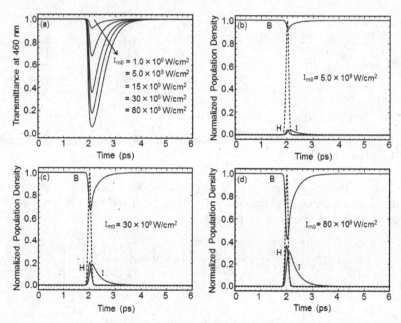

Fig. 3. (a). Variation of the transmittance of the probe laser beam with time for different peak pump intensity (I_{m0}) values at 570 nm, and (b)-(d) corresponding variation of the normalized population densities of B, H and I states with time for normalized input pulse at 570 nm *(dashed line)*

The percentage modulation increases with increase in I_{m0}, as it results in increase in the number of excited molecules in I_{460} state from the initial B state (Fig. 3(b)–(d)), resulting in increased absorption of the probe beam at 460 nm and hence decrease in its transmission. For $I_{m0} = 1.0$ GW/cm^2 and 80 GW/cm^2, the probe beam is modulated by 8.5% and 94.5% respectively, with the switch-off time of 0.2 and 0.25 ps and switch-on time of 0.9 and 2.1 ps respectively. Theoretical results are in good agreement with earlier reported experimental results [23,35,36]

The lifetime of I_{460} state has been shown to vary from 50 fs to more than 500 fs by various techniques [35,36]. Biesso et al. have reported an increase in the life time of the I_{460} state to 800 fs by varying the concentration and size of gold nanoparticles due to the plasmonic field effect [35]. Cheng et al., have recently shown that in the presence of ionic surfactants cetrimonium bromide (CTAB) and sodium dodecyl sulfate (SDS), the lifetime of the reactive excited state I_{460} increases by upto 20% [36]. The effect of I_{460} state life time τ_2 on the switching characteristics is shown in Fig. 4 (a). An increase in τ_2 results in increase in switch off/on time and switching contrast due to slower relaxation and increased population of the I_{460} state. The effect of the variation of pump pulse width (Δt) on switching characteristics is shown in Fig. 4(b).

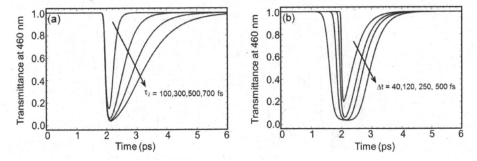

Fig. 4. Variation of transmittance of the probe laser beam with time for different (a) τ_2 values, and (b) Δt values, at $I_{m0} = 80$ GW/cm^2

In this case also, the switch off/on time and percentage modulation increase with increase in Δt, due to increased interaction time of the pulsed beam with bR molecules, with the contrast saturating beyond 250 fs. For instance, for $\Delta t = 40$ fs and 500 fs, the switch-off time is 0.13 and 0.85 ps whereas, switch-on time is 1.82 and 2.4 ps with 80% and 96% modulation respectively. As expected, the switching characteristics become symmetric for smaller τ_2 values ($\tau_2 < \Delta t$). The effect of absorption cross-section values of I_{460} state at probe wavelength (σ_{lp}) on the switching dynamics is shown in Fig. 5. The percentage modulation increases on increase in σ_{lp} as it also leads to increased absorption of the probe at 460 nm. Nearly complete switching i.e. ~100% modulation can be achieved at $\sigma_{lp} = 2.4 \times 10^{-16}$ as shown in Fig. 5.

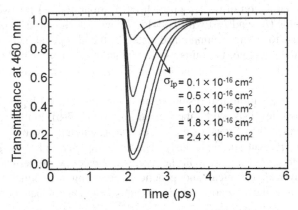

Fig. 5. Transmittance of probe beam with time for different values of σ_{lp} at $I_{m0} = 80$ GW/cm^2

The effect of the pump pulse frequency on the switching characteristics is shown in Fig. 6. The optimum pump pulse separation is 1.7 ps at I_{m0} = 80 GW/cm^2, maintaining a switching contrast of ~95% that results in a bit rate of 0.58 Tbits/sec. As mentioned earlier, bR exhibits exceptional photo and thermal stability. A number of experimental investigations have demonstrated repeated switching of bR without any photo degradation, although at lower frequencies [10,18,37]. Recent studies on coherent control of isomerization of retinal in bR in the high intensity regime have used intensities as high as ~ 200 GW/cm^2 [38].

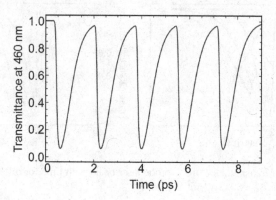

Fig. 6. Effect of pulse frequency on the transmittance of the probe laser beam at 460 nm

The switching characteristics have been used to theoretically design all-optical NOT and the universal NAND and NOR logic gates with multiple pulsed pump laser beams. For this, the rate equations have been solved by considering two pulsed pump laser beams I_{m1} and I_{m2} given by Eqs (5) and (6). The percentage modulation increases with increase in peak pumping intensity and saturates after a certain value. Hence, the difference between the normalized output transmitted intensity for single and double input pulses decreases for NAND logic gates. There is thus an optimum value of the peak pump intensity for which the difference in the two input pulses and the threshold intensity level is maximum.

Amplitude modulation of the cw probe laser beam at 460 nm for two input pulsed pump laser beams (I_{m1} and I_{m2}) with I_{m0} =15 GW/cm^2 is shown in Fig. 7(a)–(d). The inverted response to an input pump pulse results in the all-optical NOT logic gate. For all-optical NOR logic gate, the output is low when either one or both the pulses are present and is high when none of the two pulses is present, as shown in Fig. 7(a),(c) and (d).

The same configuration also results in an all-optical NAND logic gate, by considering a threshold level shown by the dashed line in Fig. 7(b). In this case, the output is high when either one or none of the pulses are present and low only when both the input pulses are present simultaneously. The switching time of these gates with the typical parameters used for bR is in the ps range. Since the properties of bR can be tailored by physical, chemical and genetic engineering techniques, the switching characteristics and the operation of the ultrafast logic gates can be optimized for

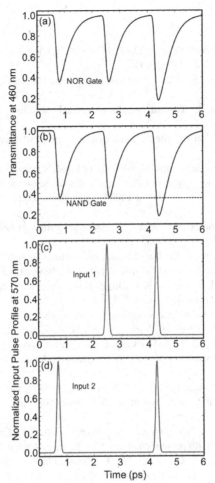

Fig. 7. All-optical logic operation (a) optical NOR gate function (b) optical NAND gate function (dashed line as the threshold level), both with the variation of normalized transmitted intensity of the probe laser beam at 460 nm as output with time; (c) and (d) are normalized input pulse profiles of the two inputs at 570 nm.

desired application. The results would be useful for all-optical ultrafast signal processing and computing due to simple digital operation.

4   Conclusion

The unique properties of bR have inspired a variety of applications in information processing. However speed has been the main limitation in its practical applicability. The present analysis demonstrates the applicability of bR for all-optical ultrafast operations in the simple pump-probe geometry and opens up exciting prospects for its use in optical supercomputing due to sub-ps switch off/on time and high switching contrast at relatively lower pump powers.

References

1. Roy, S.: Editorial, Special Issue on Optical Computing Circuits, Devices and Systems. IET Circ., Dev. and Syst. 5, 73–75 (2011)
2. Caulfield, H.J., Dolev, S.: Why future supercomputing requires optics? Nature Photon. 4, 261–263 (2010)
3. Haque, S.A., Nelson, J.: Toward organic all-optical switching. Science 327, 1466–1467 (2010)
4. Szacilowski, K.: Digital information processing in molecular systems. Chem. Rev. 108, 3481–3548 (2008)
5. DeSilva, A.P.: Molecular Computing: A layer of logic. Nature 454, 417–418 (2008)
6. Roy, S., Yadav, C.: All-optical ultrafast logic gates based on saturable to reverse saturable absorption transition in CuPc-doped PMMA thin films. Opt. Commun. 284, 4435–4440 (2011)
7. Hampp, N.: Bacteriorhodopsin as a photochromic retinal protein for optical memories. Chem. Rev. 100, 1755–1776 (2000)
8. Lukashev, E.P., Robertson, B.: Bacteriorhodopsin retains its light-induced proton pumping function after being heated to 140^0C, Bioelectrochem. Bioenerg. 37, 157–160 (1995)
9. Stuart, J.A., Mercy, D.L., Wise, K.J., Birge, R.R.: Volumetric optical memory based on bacteriorhodopsin. Synth. Metals 127, 3–15 (2002)
10. Singh, C.P., Roy, S.: All-optical switching in bacteriorhodopsin based on M state dynamics and its application to photonic logic gates. Optics Commun. 218, 55–66 (2003)
11. Chen, G., Lu, W., Xu, X., Tian, J., Zhang, C.: All-optical time-delay switch based on grating buildup time of two-wave mixing in a bacteriorhodopsin film. Appl. Opt. 48, 5205–5211 (2009)
12. Rao, D.V.G.L.N., Aranda, F.J., Rao, D.N., Chen, Z., Akkara, J.A., Kaplan, D.L., Nakashima, M.: All-optical logic gates with bacteriorhodopsin films. Opt. Commun. 127, 193–199 (1996)
13. Joseph, J., Aranda, F.J., Rao, D.V.G.L.N., DeCristofano, B.S.: Optical computing and information processing with protein complex. Opt. Memory Neural Netw. 6, 275–285 (1997)
14. Zhang, T., Zhang, C., Fu, G., Li, Y., Gu, L., Zhang, G., Song, Q.W., Parsons, B., Birge, R.R.: All-optical logic gates using bacteriorhodopsin films. Opt. Eng. 39, 527–534 (2000)
15. Li, Y., Sun, Q., Tian, J., Zhang, G.: Optical boolean logic gates based on degenerate multi-wave mixing in bR films. Opt. Mater. 23, 285–288 (2003)
16. Roy, S., Singh, C.P., Reddy, K.P.J.: Generalized model for all-optical light modulation in bacteriorhodopsin. J. Appl. Phys. 90, 3679–3689 (2001)
17. Sharma, P., Roy, S.: Effect of probe beam intensity on all-optical switching based on excited-state absorption. Opt. Mat. Exp. 2, 548–565 (2012)
18. Huang, Y., Wu, S., Zhao, Y.: All-optical switching characteristics in bacteriorhodopsin and its applications in integrated optics. Opt. Exp. 12, 895–906 (2004)
19. Roy, S., Prasad, M., Topołancik, J., Vollmer, F.: All-optical switching with bacteriorhodopsin protein coated microcavities and its application to low power computing circuits. J. Appl. Phys. 107, 053115-1–053115-9 (2010)
20. Roy, S., Sethi, P., Topolancik, J., Vollmer, F.: All-optical reversible logic gates with optically controlled bacteriorhodopsin protein-coated microresonators. Adv. Opt. Technol. 2012, 727206-12 (2012)
21. Der, A., Valkai, S., Fabian, L., Ormos, P., Ramsden, J.J., Wolff, E.K.: Integrated optical switching based on the protein bacteriorhodopsin. Photochem. Photobio. 83, 393–396 (2007)

22. Fabian, L., Wolff, E.K., Oroszi, L., Ormos, P., Der, A.: Fast integrated optical switching by the protein bacteriorhodopsin. Appl. Phys. Lett. 97, 0233051-3 (2010)
23. Petrich, J.W., Breton, J., Martin, J.L., Antonetti, A.: Femtosecond absorption spectroscopy of light-adapted and dark-adapted bacteriorhodopsin. Chem. Phys. Lett. 137, 369–375 (1987)
24. Mathies, R.A., Cruz, C.H.B., Pollard, W.T.: Direct observation of the femtosecond excited-state *cis-trans* isomerization in bacteriorhodopsin. Science 240, 777–779 (1988)
25. Dobler, J., Zinth, W., Kaiser, W., Oesterhelt, D.: Excited-State reaction dynamics of bacteriorhodopsin studied by femtosecond spectroscopy. Chem. Phys. Lett. 144, 215–220 (1988)
26. Ye, T., Friedman, N., Gat, Y., Atkinson, G.H., Sheves, M., Ottolenghi, M., Ruhman, S.: On the nature of the primary light-induced events in bacteriorhodopsin: Ultrafast spectroscopy of native and $C_{13} = C_{14}$ locked Pigments. J. Phys. Chem. B 103, 5122–5130 (1999)
27. Aharoni, A., Hou, B., Friedman, N., Ottolenghi, M., Rousso, I., Ruhman, S., Sheves, M., Ye, T., Zhang, Q.: Non-isomerizable artificial pigments: Implications for the primary light-induced events in bacteriorhodopsin. Biochem. 66, 1210–1219 (2001)
28. Kobayashi, T., Yabushita, A., Saito, T., Ohtani, H.: Real time spectroscopy of transition states in bacteriorhodopsin during retinal isomerization. Nature 414, 531–534 (2001)
29. Kobayashi, T., Yabushita, A., Saito, T., Ohtani, H., Tsuda, M.: Sub-5 fs-real time spectroscopy of transition states in bacteriorhodopsin during retinal isomerization. Photochem. Photobio. 83, 363–368 (2007)
30. Yishi, W., Sheng, Z., Xicheng, A., Kunsheng, H., JianPing, Z.: Ultrafast isomerization dynamics of retinal in bacteriorhodopsin as revealed by femtosecond absorption spectroscopy. Chin. Sci. Bull. 53, 1972–1977 (2008)
31. Yabushita, A., Kobayashi, T.: Primary conformation change in bacteriorhodopsin on photoexcitation. Biophys. J. 96, 1447–1461 (2009)
32. Briand, J., Leonard, J., Haacke, S.: Ultrafast photo-induced reaction dynamics in bacteriorhodopsin and its Trp mutants. J. Opt. 12, 1–14 (2010)
33. Abramczyk, H.: Mechanisms of energy dissipation and ultrafast primary events in photostable systems: H-bond, excess electron, biological photoreceptors. Vibrational Spectroscopy 58, 1–11 (2012)
34. Fabian, L., Heiner, Z., Mero, M., Kiss, M., Wolff, E.K., Ormos, P., Osvay, K., Der, A.: Protein based ultrafast photonic switching. Opt. Exp. 19, 18861–18870 (2011)
35. Biesso, A., Qian, W., Sayed, M.: Gold nanoparticle plasmonic field effect on the primary step of the other photosynthetic system in nature, bacteriorhodospin. J. Am. Chem. Soc. 130, 3258–3259 (2007)
36. Cheng, C., Lee, Y., Chu, L.: Study of the reactive excited-state dynamics of delipidated bacteriorhodopsin upon surfactants treatments. Chem. Phys. Lett. 539-540, 151–156 (2012)
37. Wu, P., Rao, D.V.G.L.N., Kimball, B.R., Nakashima, M., DeCristofano, B.S.: Enhancement of photoinduced anisotropy and all-optical switching in bacteriorhodopsin films. Appl. Phys. Lett. 81, 3888–3890 (2002)
38. Prokhorenko, V., Halpin, A., Johnson, P., Miller, R., Brown, L.: Coherent control of the isomerization of retinal in bacteriorhodopsin in the high intensity regime. J. Chem. Phys. 134, 085105(1–5) (2011)

A Nano-Optics Vector Matrix Multiplier for Implementing State Machines*

Eyal Cohen[1], Shlomi Dolev[1] and Michael Rosenblit[2]

[1] Department of Computer Science
Ben Gurion University of the Negev, Israel
[2] Ilze Katz Institute for Nanoscale Science & Technology
Ben Gurion University of the Negev, Israel
{dolev,eyalco}@cs.bgu.ac.il, rmichael@bgu.ac.il

Abstract. We present a design concept for a nano optical architecture for a finite state machine. The architecture uses the Vector-Matrix-Multiplier as the basic device used to perform calculations. We provide schematics of such a device. These devices may in turn provide the building blocks for optically controlled devices.

1 Introduction

Optical technologies are gradually taking over the computer world. Optical data storage such as CDs and DVDs, and optical communication such as fiber optics are proving more efficient than their electronic counterparts. An effort towards computing using optical elements is done to suggest novel computing devices.

Computer scientists use Turing Machines to model the computer and discuss its operation. However, the common Von Neuman architecture is different than the Turing machine. The Turing Machine has an infinite tape as its memory while the size of present day computers memory is finite. The Von Neuman architecture could be modelled by the Finite State Machine [1], though the overhead of using this model would be too high because of the large size of the memory. It is preferred to model the Von Neuman architecture with small units of processing similar to finite state machines, controlled by a program.

An "Optical Finite State Machine" can be a major step towards all optical computers. The Vector-Matrix-Multiplier [2,3] was suggested as a building block for devices designed to solve specific computation tasks. We propose building Optical Finite State Machines based on the Vector-Matrix-Multiplier.

* Partially supported by a Russian Israeli grant from the Israeli Ministry of Science and Technology and the Russian Foundation for Basic Research, the Rita Altura Trust Chair in Computer Sciences, the Lynne and William Frankel Center for Computer Sciences, Israel Science Foundation (grant number 428/11), Cabarnit Cyber Security MAGNET Consortium, Grant from the Institute for Future Defense Technologies Research named for the Medvedi of the Technion, MAFAT, and Israeli Internet Association.

S. Dolev and M. Oltean (Eds.): OSC 2012, LNCS 7715, pp. 78–91, 2013.

Section 1 presents an introduction to the proposed device. We present the optical building blocks which are currently available for usage. We review the theoretic background of the finite state machine, and we suggest the state machine using the vector matrix multiplication method. Section 2 presents the details of the presented device. We present the design of the optical elements to be used, and the overall design of the device. We conclude in section 3 with the future work directives that will allow us to build the device.

1.1 Building Blocks

Vector Matrix Multipliers. An optical Vector Matrix Multiplier (VMM) projects an array of light over a mask and then recollects the light in order to calculate an output vector. The mask encodes the vector which we would like to multiply with. The output vector could be collected back to fiber optics or read by an array of detectors. Perhaps the most famous setup that performs this operation is the Stanford VMM [2,3]. An optical setup broadens the input vector to be projected over the entire matrix such that each input element is projected over a respective row. The multiplication occurs when light is masked by the matrix. Then the optical setup collects the multiplied light such that each column is collected and projected on a respective output column element. It has been suggested that the VMM should be used for creation of coprocessors [3] or used as the main processing tool on a processor [4].

Nano Antennas. *Nano antennas* are devices that reshape the propagation profile of light as a result of interaction between the antenna's elements to the electromagnetic radiation in space. It has been suggested that the plasmon resonance of metal nano particles can direct light from optical emitters. In order to implement the VMM we suggest using nano antennas.

Nano antennas are similar to normal antennas by utilizing a geometric array of elements to manipulate the interaction of a dipole element with the electromagnetic radiation. This could be done to amplify emission towards a certain direction and polarity or to amplify the reception of a signal coming from a certain direction. The size of the antenna's elements should be smaller than quarter of the wavelength. This presents no problem in the radio-frequency wavelengths by choosing metal elements ranging from a few centimeters to a few meters. However nano technology design is an essential part of building antennas for the optical band.

Some designs of the nano antennas borrow their design from the radio-frequency regime for example the nano optical Yagi-Uda antenna [5]. The nano optical Yagi-Uda antenna design suggests using nano metallic elements. Instead of driving the dipole with an electron current, the dipole is driven by an optical wave. The dipole is tilted in 45 degrees compared to the other elements. A polarized light is projected over the dipole while the polarization would allow coupling of the radiation only to the dipole. The dipole and the other elements coupled to it would now emit light in another direction and polarity.

Other nano antennas present completely new designs. For example in [6], the design suggests the coupling of a big dielectric sphere (with a diameter of $500nm$) to two metallic particles (with a diameter of $60nm$). This design exhibits a directional radiation profile and amplification for the emission efficiency.

We suggest using such nano antennas as the mechanism to collect, reflect and emit light to and from the elements of the VMM. Devices such as the nano optical Yagi-Uda antennas could redirect light of a certain polarity and direction towards a different direction. A selection mechanism can be chosen that may allow us to deactivate some of the antennas by an optical signal. This may be done by choosing material for the elements that would stop interacting over a certain threshold. The selection mechanism should only add enough optical intensity that would throw the element over or below the threshold.

Another selection mechanism could utilize the antennas original property as directional device to manipulate two beams of light. When two beams of light propagate towards the antenna, the antenna reflects a beam of light into a certain direction and when only one of those beams propagates, the light projects towards another direction.

1.2 Finite State Machine

A finite state machine is an abstract machine that can be in one of a finite number of states [1]. It is defined by a transition table where on one side of the table, we have the current state of the machine, and the input, and on the other side we have the next state of the machine. The state and the input are defined by a finite set of bits. For simplicity we start considering binary bits as opposed to using shades of gray. We assume that the state is represented by n bits and the input data is represented by m bits. Therefore the output, which must represent the next state, should have at least n bits together with an optional additional k bits output. A general transition table should have the form of Table 1.

Where the q vector represents a state, the d vector represents an input, the Q vector represents the next state and the D vector represents additional output. We can say that the current state q together with the input d represents a bigger global input which we refer to as the v vector, where:

$$q_i = v_i$$

$$d_i = v_{i+n}$$

and the overall size of v is $n + m$. In the same manner we can say that the next state Q together with the output D represents a bigger global output which we refer to as the u vector, where:

$$Q_i = u_i$$

$$D_i = u_{i+n}$$

and the overall size of u is $n + k$.

For example we use the finite state machine identical to the one presented in [1]. The transition diagram is represented in Figure 1. Its transition table is represented in

Table 1. Typical transition table

Current state	Input	Next state	Add. Output
$q_n \ldots q_3\, q_2\, q_1$	$d_m \ldots d_3\, d_2\, d_1$	$Q_n \ldots Q_3\, Q_2\, Q_1$	$D_k \ldots D_3\, D_2\, D_1$
0 ... 0 0 0	0 ... 0 0 0	0 ... 0 0 0	0 ... 0 1 0
0 ... 0 0 0	0 ... 0 0 1	0 ... 0 1 0	0 ... 1 0 1
0 ... 0 0 0	0 ... 0 1 0	0 ... 0 0 1	1 ... 0 1 0
$\vdots$	$\vdots$	$\vdots$	$\vdots$
0 ... 0 0 0	1 ... 1 1 1	0 ... 1 0 1	0 ... 1 0 1
0 ... 0 0 1	0 ... 0 0 0	0 ... 0 0 1	1 ... 0 0 1
0 ... 0 0 1	0 ... 0 0 1	0 ... 1 0 0	1 ... 0 0 0
0 ... 0 0 1	0 ... 0 1 0	0 ... 0 1 0	0 ... 0 1 0
$\vdots$	$\vdots$	$\vdots$	$\vdots$
0 ... 0 0 1	1 ... 1 1 1	0 ... 1 1 1	1 ... 1 0 1
0 ... 0 1 0	0 ... 0 0 0	0 ... 1 0 1	0 ... 1 0 1
$\vdots$	$\vdots$	$\vdots$	$\vdots$
1 ... 1 1 1	1 ... 1 1 1	1 ... 1 0 1	0 ... 1 0 1

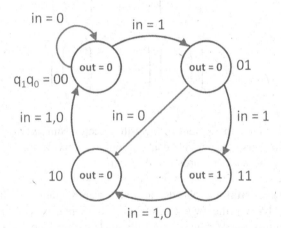

Fig. 1. A finite state machine states diagram. According to a state machine given in [1].

Table 2. In our example the current state is represented by two bits, the input by one bit, the next state by two bits and the output by one bit. Thus in this case the global input (the v vector) is represented by three bits, and the global output is represented three bits.

1.3   Finite State Machine Using Matrix Multiplication

Given a finite state machine, it is possible to represent its transition table as a transformation matrix M where multiplying it by a given v vector would result in an appropriate

Table 2. Transition table of the state machine in Figure 1

Current state		Next state	
$q_1\ q_0$	Input	$Q_1\ Q_0$	Output
00	0	00	0
00	1	01	0
01	0	10	0
01	1	11	0
10	0	00	0
10	1	00	0
11	0	10	1
11	1	10	1

u vector. We need to calculate the elements of the matrix M. To do so we need to define matrices V and U by arranging the v and u vectors vertically one by one in the V and U matrices respectively.

$$(V) = \left(\begin{pmatrix} \vdots \\ v_1 \\ \vdots \end{pmatrix} \begin{pmatrix} \vdots \\ v_2 \\ \vdots \end{pmatrix} \cdots \begin{pmatrix} \vdots \\ v_l \\ \vdots \end{pmatrix} \right)$$

$$(U) = \left(\begin{pmatrix} \vdots \\ u_1 \\ \vdots \end{pmatrix} \begin{pmatrix} \vdots \\ u_2 \\ \vdots \end{pmatrix} \cdots \begin{pmatrix} \vdots \\ u_l \\ \vdots \end{pmatrix} \right)$$

Thus V_{ij} represents the i-th element of the j-th v vector. Similarly U_{ij} represents the i-th element of the j-th u vector. We want to find a matrix M that would satisfy the equation $MV = U$ or in other words: $M_{ij}V_{jk} = U_{ik}$

Using Unary Representation. We use an unary representation of the v vectors by using the standard basis as the v vectors, such that the V matrix is a unity matrix. Thus if the V matrix is unity, the M matrix is actually the U matrix. The M matrix is a matrix of zeros and ones such that each column has one element that equals 1 while the others are 0 (zeros). This way, when multiplying a basis vector by that matrix we get another basis vector which represents a new state and output.

For example, using the table above, we can translate the binary representation of the states and inputs and outputs to the decimal representation to derive the values of Table 3.

The following matrix represents the transformation matrix that would transform the state and the input vector to the next state and output vector,

Table 3. Transition table in decimal representation

State and input	Next state and output
0	0
1	2
2	4
3	6
4	0
5	0
6	5
7	5

$$M = \begin{pmatrix} 1\,0\,0\,0\,1\,1\,0\,0 \\ 0\,0\,0\,0\,0\,0\,0\,0 \\ 0\,1\,0\,0\,0\,0\,0\,0 \\ 0\,0\,0\,0\,0\,0\,0\,0 \\ 0\,0\,1\,0\,0\,0\,0\,0 \\ 0\,0\,0\,0\,0\,0\,1\,1 \\ 0\,0\,0\,1\,0\,0\,0\,0 \\ 0\,0\,0\,0\,0\,0\,0\,0 \end{pmatrix},$$

where each column represents the next state and output, whose values are selected by the position of the single 1 in the column.

Unary Representation Back to State and Output. Using this unary representation is possible only if we can transform the input and state to the unary representation. Similarly transforming the output unary vector to a new state vector and an output.

The latter is derived very easily. Note that the vectors in V form the normal basis thus a matrix where each column represents the output and the new state can be used. Transforming an input vector and state vector to a single unary vector is somewhat more difficult. First assume that those two vectors are unary too.

We can extend these two vectors and use a bitwise logical "and" operation on all their bits as follows. Assuming that the state vector is sized n where the bit numbered i equals 1 and all other bits are 0. In addition, assume that the input state vector is sized m where the element numbered j equals 1 and all other elements are 0. We would like to create a vector sized $n \cdot m$ with the $m \cdot i + j$ bit equals 1 and all other elements are 0.

The binary code for the state and the binary code for the input are extended in different ways. The binary code of the state is extended by duplicating each individual bit m times, such that all bits are zero except the bits with indexes greater or equal $m \cdot i$ and smaller or equal $(m + 1) \cdot i - 1$. Recall that the bit i was the single bit with the value 1 in the original binary representation of the state.

The binary code for the input is replicated n times as a monolithic block (When considering each individual bit). In the resulting extended input, every two bits with index difference of m have the same value.

For example assume the following vectors:

$$
\begin{pmatrix} 0 \\ 1 \\ 1 \\ 0 \end{pmatrix}, \begin{pmatrix} 0 \\ 0 \\ 1 \\ 0 \end{pmatrix},
$$

We can extend them as follows:

$$
\begin{pmatrix} 0 \\ 1 \\ 0 \\ 0 \\ 1 \\ 0 \\ 0 \\ 1 \\ 0 \\ 0 \\ 1 \\ 0 \end{pmatrix} \times \begin{pmatrix} 0 \\ 0 \\ 0 \\ 0 \\ 0 \\ 0 \\ 1 \\ 1 \\ 1 \\ 0 \\ 0 \\ 0 \end{pmatrix} = \begin{pmatrix} 0 \\ 0 \\ 0 \\ 0 \\ 0 \\ 0 \\ 0 \\ 1 \\ 0 \\ 0 \\ 0 \\ 0 \end{pmatrix}.
$$

Multiplying or using bitwise logical "and" operation we can get the required result. This multiplication could be done by projecting these two vectors over an array of detectors and using a threshold level. This level should be above the intensity of one beam and under the intensity of two beams projected together.

2   Optical Finite State Machine

We present a device that implements finite state machines using optics, and therefore can be a step towards all-optical-processor. We suggest an implementation of the optical finite state machine using the Vector-Matrix-Multiplication technique. In order to use mostly optical technology, the programming and processed data are written and read to and from an optical memory [7,8].

Many designs of Vector-Matrix-Multipliers were suggested and none of which were nanoscaled. The architecture we chose to implement with nano scale devices appears on Figure 2. The input vector emits light in such a manner that each pixel emits a different light intensity. This light hits the multiplication matrix, and is reflected towards the output vector. Each pixel of the multiplication matrix has a different reflection coefficient, this way the reflected light has a value of an input intensity multiplied by a reflection coefficient. Light propagating from each row of the input vector should hit an appropriate pixel of the multiplication matrix. Thus the first element of the input vector is projected over the first column of the matrix where it will be multiplied by the reflection coefficient, when it is reflected over the first output vector element. Similarly the n-th pixel is projected over the n-th column, multiplied and reflected over the n-th output vector element.

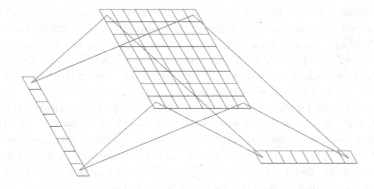

Fig. 2. Schematic design of the Vector-Matrix-Multiplier

After the reflection off the multiplication matrix, the obtained beams intensities are collected and directed to the output vector, summing the intensities over multiplied beams. The summing process is done by either using a lens or by designing the reflective multiplication matrix to act as a concaved mirror, reflecting the appropriate beams in a proper direction. Each column of the input matrix should be summed after the multiplication over an appropriate pixel of the output matrix. The first column of the input matrix should be summed over to the first pixel of the output vector after the multiplication. The n-th column of the input matrix should be summed over to the n-th pixel of the output vector after the multiplication.

Notice that the optical elements which represent the arrays of matrix and vectors are arranged in layers. This makes it possible to build this device over the layers of a substrate similarly to current processor production technology. This process is similar to the planar optics technology presented in [9].

Reading and Writing to Memory. Special attention should be given to the way that data is read and written over the memory. We would like to be able to control the input matrix and the multiplication vector, such that their coefficient values would be determined by values stored over an optical memory. This could be done by one of two ways: (a) An optical memory will emit light which will control the appropriate coefficients. (b) The input matrix and the multiplication vector would be an optical memory device by themselves, controlled by light emitted by other devices or reflected back from this device's output.

The memory should keep the data valid as long as it takes to switch to another state. If the steps of the state machine are controlled by a clock, then the memory should hold the data as long as the period of this clock.

Such memory device may be connected to an electrical input to power the device, or amplify the propagating light but it should be controlled only by an optical input. The memory devices may use different methods such as cavities, memristors, or second harmonic generation materials. For example in [7] the use of ring resonator is suggested. A high enough intensity provided by the bus waveguide would electrically charge the semiconductor substrate. This charge caused a minute change in the refraction index. This minute change creates a major effect on the transmittance for a certain wavelength. Using a reading system from an array of such ring resonators in that wavelength is suggested. The only problem with this device is that it has to be electrically erased (EEPROM). This may not have to present a problem if the erasing process is very short as the period of the clock chosen for the device.

A different optical memory device is suggested in [8]. This device suggests interaction of waveguides in a crystal of nanocavities. The geometry of the device makes the possibility to achieve bi-stable states of the outputs of the device. When going over a certain threshold of the input power, the system change into one state, and when going lower than another threshold the system changes into another state. There is a range between those thresholds which simply keeps the current state.

2.1   Implementation Using Optical Devices

We would like to create a specific state machine which will be chosen according to the limits of the design. As the automata becomes more complex with more states and bigger input, a much bigger vector and matrices should be used.

We assume that the state is given in an unary state form, and also the input is given unary form. The value is determined optically where only one of the pixels or elements in a channel propagates light while the other elements do not carry light propagating at the same time.

Below we present the optical elements we use to build the optical device.

Vector Extension Mechanism. When an input vector, and a state vector are given, they should combine together into a bigger global input. As described earlier we would like to extend the two vectors. We should widen each one of the pixels on one vector, while making copies of the other vector a few times. This may be done by using passive devices like lens or semi transparent mirrors. Schematics of these processes done by mirrors could be seen on figures 4 and 5. The widening and the duplication systems could be placed as layers one on top of the other. The extended vector are created in two layers and are ready to be passed to the "and" mechanism.

These extended vectors should preform an "and" operation, that is using an active mechanism to allow only areas where both pixels contain light while darkening pixels that do not contain light, or contain light on only one of the channels. The global input chosen by the vector extension mechanism contains an unary vector, with only one pixel containing light, where the values of both extended vectors are on. The schematics of the "and" mechanism is described on figure 6. The schematics accepts two extended vectors and create one global input.

In order to implement this device we need an element that performs the logic "and". The idea is to design an element that allows light to propagate only when the intensity grows beyond a certain value. Other designs simply suggests that one signal controls the other [10]. By doing so it is possible to achieve the logic "and". Many devices which utilize those approaches are using ring or disc resonators. The ring resonators provide the feedback loop needed to intensify the nonlinear effect to achieve a bi-stable state.

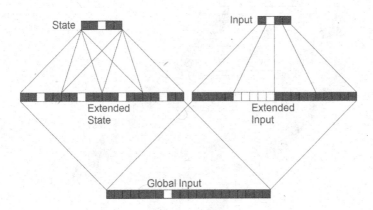

Fig. 3. Design of the extension mechanism. The global input vector is created by duplicating the the state vector and extendin each element of the input vector. Next, an "and" operation is performed.

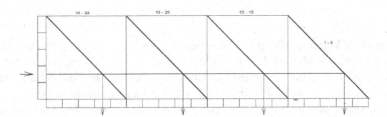

Fig. 4. Schematics of the duplication mechanism used to duplicate the state vector. The unaric state vector is propagating through the device and being reflected over the appropriate ports.

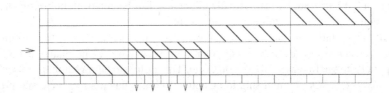

Fig. 5. Schematics of the widening mechanism used to widen each element of the input vector. The unaric input vector is propagating through the device and being reflected over the appropriate ports.

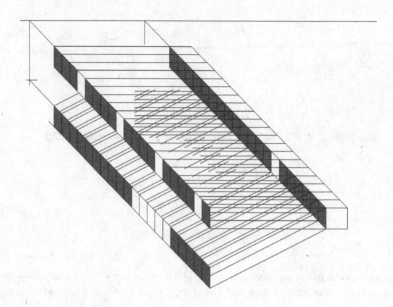

Fig. 6. Schematics of the "and" mechanism. Both vectors are projected over an active device which allows signal over a certain threshold. The threshold is determined by the power of two light signals.

Multiplication Matrix. The global input can now act as a selection mechanism together with a Vector-Matrix-Multiplier. Only one of the pixels of the vector contains light. When projecting this vector over a matrix, in effect this one pixel chooses one of the columns of the matrix by lighting it. The vector created contains the next state and the output. This vector is split where some of its elements contain the next state, while the other elements contain the output. The next state is again an unary vector to be used later, and the output could be unary or binary again because the system does not need to use it again. The design uses the light of the global input created in the previous step. Figure 7 illustrates the logic of the selection mechanism using the VMM.

A schematic design of the VMM is given earlier on figure 2. This design features flat elements which form parallel layers. This makes it possible to build this element on the surface of a flat substrate by utilizing 2D elements. The beams of light propagate in a 3D environment due to the width of the substrate.

The implementation of the state machine limits some parameters of the design. By choosing the matrix we are actually choosing a state machine. The matrix is constant and does not change for each calculation cycle. The multiplication vector is unaric. This means only one vector element should be active for each calculation cycle and project light over the matrix.

The design scheme utilizes exact projection, reflection and collection of light by the matrix elements. Normal mirrors, projectors and other optical elements are very efficient in scales larger than the wavelength. However, they may suffer from poor

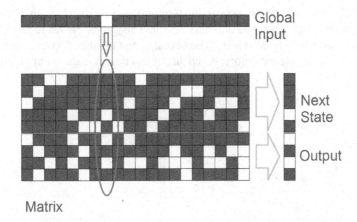

Fig. 7. Design of the Selection mechanism. The global input chooses a certain vector of the matrix, which holds the next state and the output.

precision and efficiency when trying to work in the nanometric scale. We suggest using nano antennas together with other nano structures as the elements of the VMM to be used in the nanometric scale.

The vector elements should project light over the matrix in a manner that each vector element would be projected over only one column of the matrix. Nano antennas with wide transmission aperture in one direction, and very precise and narrow aperture in the other direction should be chosen. Similarly each output vector should use nano antennas with wide aperture in one direction an very small and precise in the other direction.

The matrix elements which reflect the light could also be built by using nano antennas. Each element should receive light projected from exactly one element of the multiplication vector and transmit it towards a single element of the output vector. Therefore each element of the matrix should hold two antennas for reception and transmission where the former would transfer the light received to the transmission antenna. Each element on the matrix is different where the antenna elements are directed towards different directions.

The truth table of the state machine defines the existence of those antenna elements. We should begin with a constant matrix such that only where the truth table suggests, we can add or remove the antennas. A more complex design would implement a programmable matrix. This could be done if we could control each element. The idea is to cut the connection between the two antennas and allow their connection only if a beam of light is projected from another direction such as from the other side of the substrate.

2.2 Device Design Concept

The elements described in the previous section are the building blocks of the device. The input is given as an unary vector and extended by the extension mechanism. Similarly, an unary state is provided by the state memory, and extended by the extension mechanism. These two vectors are passing through the "and" mechanism, and the result is the global input presented as an unary vector. This vector passes to the VMM and chooses an output and the next state. The output is sent to other devices, while the next state is stored on the state memory. A scheme of this device is shown in Figure 8.

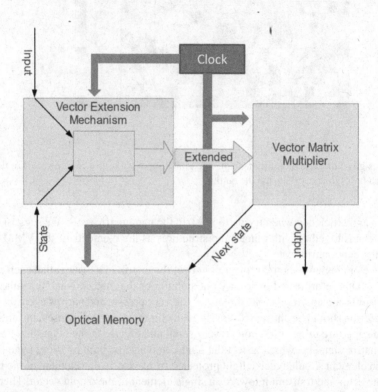

Fig. 8. Scheme of the optical device. The Vector Matrix Multiplier is used as the main building block together with the Vector Extension Mechanism and the Optical Memory, as described in 2.1.

3   Future Work

Calculations should be done in order to choose the numerical properties of each element discussed above. The properties should fit the wavelength and materials that will be chosen. Later, a simulation should be done to identify the limits of the design.

As described earlier it is also suggested to use a matrix that is not constant and which is in fact programmable. This could be done by controlling the antennas of each element by enabling or disabling it according to another optical input.

Another possibility for enhancing the device could utilize frequency multiplexing. We can use different frequencies to perform different calculations. It is possible that after each multiplication, the state bits get a frequency shift, hence moving to another color. The state is now combined with a new input and then multiplied again with the same matrix. This color may interact differently with the matrix, namely a different multiplication will be applied to each color. This is an equivalent of multiplication with another matrix on the same device.

The input in the different colors could be processed simultaneously. This could increase the computing speed in a great factor depending on the multiplexing resolution. The system could also wait until the next input arrives in a different colour. Since there are no interferences between the different colors, the system simply waits until the next colour comes into play.

References

1. Ward, S.A., Halstead Jr., R.H.: Computation structures. MIT electrical engineering and computer science series. MIT Press (1990)
2. Goodman, J.W.: Introduction to Fourier optics. McGraw-Hill, New York (1996)
3. Goodman, J.W., Dias, A.R., Woody, L.M.: Fully parallel, high-speed incoherent optical method for performing discrete fourier transforms. Optics Letters 2(1), 1–3 (1978)
4. Avner Goren, S., Sariel, A., Levit, S., Asaf, Y., Liberman, S., Sender, B., Tzelnick, T., Hefetz, Y., Moses, E., Machal, V.: Vector-matrix multiplication (May 2009)
5. Kosako, T., Kadoya, Y., Hofmann, H.F.: Directional control of light by a nano-optical yagiuda antenna. Nature Photonics 4, 312–315 (2010)
6. Krasnok, A.E., Miroshnichenko, A.E., Belov, P.A., Kivshar, Y.S.: All-dielectric optical nanoantennas. arXiv:1206.5597v1 [physics.optics] (2012)
7. Barrios, C.A., Lipson, M.: Silicon photonic read-only memory. Journal of Lightwave Technology 24(7), 2898–2905 (2006)
8. Notomi, M., Tanabe, T., Shinya, A., Kuramochi, E., Taniyama, H.: On-chip all-optical switching and memory by silicon photonic crystal nanocavities. In: Advances in Optical Technologies (2008)
9. Sinzinger, S., Jahns, J.: Integrated micro-optical imaging system with a high interconnection capacity fabricated in planar optics. Applied Optics 37(20), 4729–4735 (1997)
10. Xu, Q., Lipson, M.: All-optical logic based on silicon micro-ring resonators. Optics Express 15(3), 924–929 (2007)

Optical Energy Efficient Asynchronous Automata and Circuits

Amir Anter[1,*], Shlomi Dolev[1], and Joseph Shamir[2]

[1] Ben-Gurion University of the Negev, Israel
{anter,dolev}@cs.bgu.ac.il
[2] Technion - Israel Institute of Technology, Haifa, Israel
jsh@ee.technion.ac.il

Abstract. An optical architecture for energy efficient asynchronous automata is suggested. We use a logical paradigm called "Directed Logic", based on the most basic reversible and energy efficient gate, the Fredkin gate. Directed Logic circuits for basic boolean gates as NOT, OR/NOR and AND/NAND are used. These circuits are then employed for an optical energy efficient automata. A D latch is then used to define the automata operation cycle. A set-reset latch is used as part of an handshake protocol that suggests an optical energy efficient automata operating internally in an asynchronous fashion. Lastly, we propose a circuit for asynchronous cascading between two automata.

Keywords: Optical Logic, Reversible Computing, Zero Energy Computing, Asynchronous Circuits, Automata.

1   Introduction

The principles of thermodynamics allow to evaluate a sequence of logic operations in zero energy [2,3,10]. If such architecture is employed, the challenge of energy consumption in electric computers can be resolved. Optical computing has existed for many years, one challenge of optical computing is solving "hard" problems faster than electronic computing [1,6]. Recently, another challenge for optical computers has appeared, namely its ability to perform calculations using zero energy [4,5]. Optical computing is suitable for the use of reversible elements, allowing zero energy calculation [7]. It was also suggested that electronic reversible elements may not be suitable to use for zero-energy calculations [5]. Hence, direct logic architecture was introduced, using optical reversible elements [8]. The aim of this work is to build a general automata based on direct logic.

---

* Partially supported by a Russian Israeli grant from the Israeli Ministry of Science and Technology and the Russian Foundation for Basic Research, the Rita Altura Trust Chair in Computer Sciences, the Lynne and William Frankel Center for Computer Sciences, Israel Science Foundation (grant number 428/11), Cabarnit Cyber Security MAGNET Consortium, Grant from the Institute for Future Defense Technologies Research named for the Medvedi of the Technion, MAFAT, and Israeli Internet Association.

S. Dolev and M. Oltean (Eds.): OSC 2012, LNCS 7715, pp. 92–104, 2013.

In addition, this automata should be used without a clock, in an asynchronously fashion. Asynchronous circuits benefit from low power consumption, high operation speed and better modularity [13]. The resulting architecture can thus lead to an optical zero-energy asynchronous Turing machine.

2   Directed Logic

We here by present a logical paradigm called "Directed Logic" that was introduced previously [8]. Directed Logic circuits are built from networks of simple elements, each performs as a simple switching operation on its input vector. The input vector (I_1, I_2) can be any of the following: $(0,0), (0,1), (1,0)$, where $(1,1)$ is not admissible. The output is also a vector (O_1, O_2). Each element performs one of two opposite operations on its input vector: pass (P) or switch (S). P is the identity operation and S is the negation operation. Thus, $S(0,1) = P(1,0) = (1,0), S(1,0) = P(0,1) = (0,1), S(0,0) = P(0,0) = (0,0)$. The choice of operation is determined by the control input of the element — C. In case $C = 0$, the elements operation will be P. If $C = 1$, the elements operation will be S. This demonstrate a simple Directed Logic element based on a Fredkin gate [7], a reversible control swap gate as presented in Figure 1.

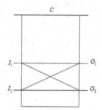

Fig. 1. Directed Logic Fredkin Gate

Several basic Directed Logic circuits are presented in in Figures 2,3 and 4, later used for the automata architecture.

The use of Fredkin gates in Directed Logic leads to reversible and conservative computational processes. In every circuit there are just as many 1 and 0 inputs as 1 and 0 outputs, hence there is no information loss. Conventional implementations of boolean logic destroy information and suffer from energy cost. In a typical boolean operation, for example NAND, there is less information at the output than at the inputs. Information loss is transformed to heat. In Directed Logic the energy consumed in practice is affected by how many output bits are lost or deleted and not by the operation of the gate, where each lost bit gives an energy loss of $kTln(2)$ [2,3,10]. One challenge, that is not covered in this paper, is to efficiently save these bits.

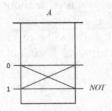

Fig. 2. NOT Circuit in Directed Logic

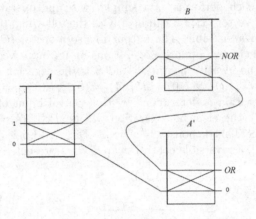

Fig. 3. OR/NOR Circuit in Directed Logic [8]

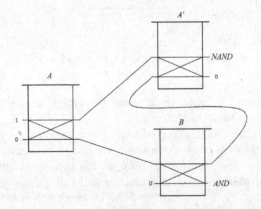

Fig. 4. AND/NAND Circuit in Directed Logic [8]

3   Optical Implementation

Directed Logic is specially adapted to the features and promises of optics. Although Directed Logic is based on Fredkin gates, there are crucial differences between them. In Fredkin's implementation, the three inputs considered as being interchangeable, hence, any output of one gate could be used as any input for a subsequent gate [7]. In Directed Logic, the controlling input in the circuit is kept entirely separate from the other two input lines. This property facilitates the optical implementation. Conservative logic can also be implemented in other ways, but Directed Logic suggest an optically implementable logic, which is hard to find. We base the architecture on possible implementations of an optical controlled switch and other advanced optical network implementations, as presented in [8]. Optical Fredkin gates were already introduced in [12]. More recent possible implementations for an optical controlled switch are Polarization switching gate [15], Acousto-optic gate, Photorefractive gate, Waveguide coupler gate and Mach-Zehnder gate.

4   2-State Energy Efficient Optical Automata

We define $2 - state$ automata by a state transition table as shown in Figure 5.

State/Input	0	1
S_0	S_i	S_l
S_1	S_j	S_k

Fig. 5. State Transition Table

The alphabet is $0, 1$ and the states are S_0, S_1, where $S_0 = 0$ and $S_1 = 1$.

Building the Automata. We define in Figure 6 a circuit later used to compute the general automata circuit. Specific S and I determine a transition to state S_n, according to the transition table and $S_n^{out} = S_n$.
We define the following:

$$S_i^{new} = S_i^{out} \wedge \neg I \wedge \neg S \tag{1}$$

$$S_j^{new} = S_j^{out} \wedge \neg I \wedge S \tag{2}$$

$$S_l^{new} = S_l^{out} \wedge I \wedge \neg S \tag{3}$$

$$S_k^{new} = S_k^{out} \wedge I \wedge S \tag{4}$$

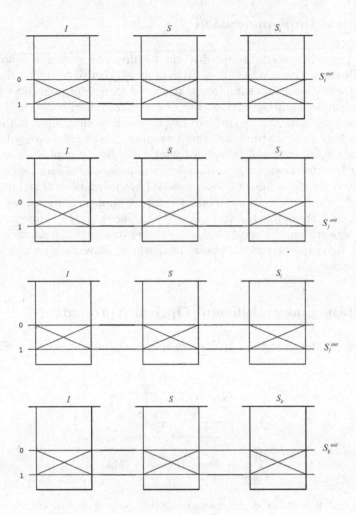

Fig. 6. Circuit to Use in Order to Compute a General Automata Circuit

Please note that for any S and I, exactly one S_n^{new} can be equal to 1 and all others S_m^{new} are 0. We define the following:

$$S^{new} = S_i^{new} \vee S_j^{new} \vee S_l^{new} \vee S_k^{new} \tag{5}$$

The output S^{new} is the new state of the automata according to the state transition table in Figure 5. We already defined the Directed Logic circuits for the OR, NOT and AND operations, now we can conclude that every gate can be implemented in Directed Logic. We define in Figure 7 a circuit for S_k^{new}.

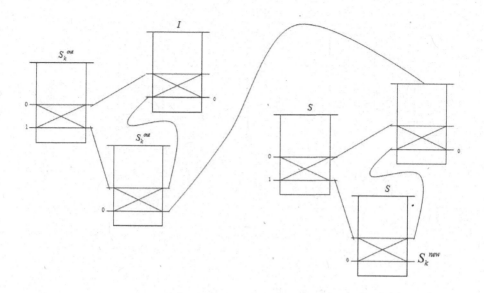

Fig. 7. Circuit of S_k^{new}

We define in Figure 8 a circuit for $S_i^{new} \vee S_j^{new} \vee S_l^{new}$ that can be used to build a circuit for S^{new}.

$2 - state$ automata based on reversible optical architecture using direct logic elements is depicted in Figure 9.

Automata Cycle. A mechanism is needed for updating the current state S of the circuit to be the S^{new} output. We use a D latch and set the input D of the latch to be the S^{new} output. Before inserting a new input I to the circuit, we set the latch, the Q output of the latch is S^{new}, we update it as the current state S of the circuit and then we reset the latch.

Figure 11 depict the traditional electrical D latch circuit.

A reversible electric D latch was introduced in [14], we use the same latch architecture for a latch composed from Directed Logic elements. We first define a NOR circuit, then we turn to define a set-reset (SR) latch, as shown in Figure 12.

We can build a gated D latch based on the SR NOR latch, as shown is Figure 13, where the S and R outputs are the inputs of the SR NOR latch. A D latch that is based on reversible optical architecture using direct logic elements, is depicted in Figure 14.

A full asynchronous cycle circuit using the D latch and the automata is shown in Figure 15.

The mechanism for handling the inputs is described in Algorithm 1.

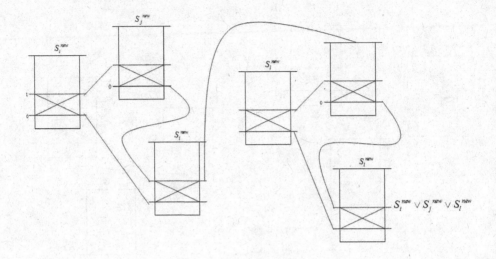

Fig. 8. Circuit for $S_i^{new} \vee S_j^{new} \vee S_l^{new}$

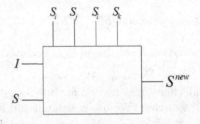

Fig. 9. 2-State Energy Efficient Optical Automata

E	D	Q	Comment
1	1	1	Set
1	0	0	Reset
0	X	Q_{prev}	No change

Fig. 10. D Latch State Transition Table

5   Asynchronous Automata

The main idea of asynchronous circuits is that different parts of the circuit are allowed to carry out their tasks at their own pace. However, in order to fulfill

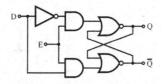

Fig. 11. A Gated electric D Latch Based on an SR NOR Latch

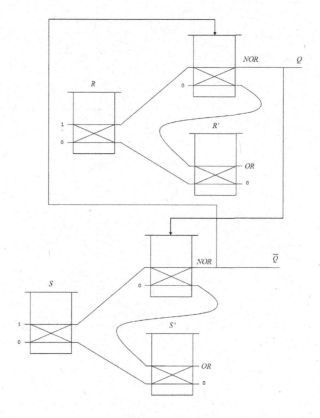

Fig. 12. SR NOR Latch

a useful purpose, the information flow between parts of the circuit must occur in a certain sequence. The main advantages of asynchronous circuits are low power consumption, high operation speed and better modularity. The output S^{new} should change according to the specific input I, according to to the state transition table of the automata. When the automata calculated the output S^{new} for some input I, the S input in the circuit should be updated to S^{new}.

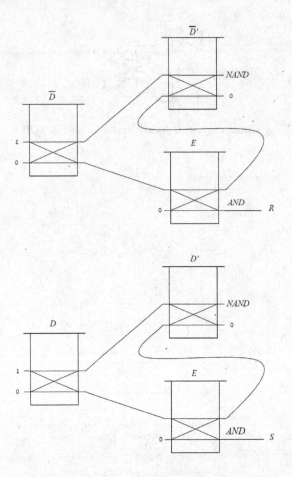

Fig. 13. A Gated D Latch Based on an SR NOR Latch

Fig. 14. Energy Efficient Optical D Latch

Then the automata should be ready to handle the next input. A clock can be used in order to control this cycle, but we would like to use an asynchronous mechanism.

Handshake Protocol. Asynchronous circuits communicate via handshakes. A handshake consists of a series of signal events sent back and forth between the

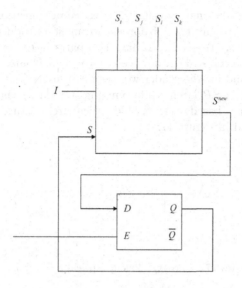

Fig. 15. 2-State Asynchronous Energy Efficient Optical Automata

---

Algorithm 1. Automata Mechanism

1: Insert the input I to the circuit.
2: The automaton operates and generates a result S^{new}
3: S^{new} is set to be the D input of the D latch.
4: The E input of the D latch is set to 1.
5: The Q output of the D latch is S^{new}.
6: S is set to be the Q output of the D latch;
7: The E input of the D latch is set to 0.
8: The automaton is ready for the next input.

S	R	Q	Comment
1	0	1	Set
0	1	0	Reset
0	0	Q_{prev}	No change

Fig. 16. D Latch State Transition Table

communicating elements [9,13]. We can divide communicating elements into a sender and a receiver. If the sender wants the receiver to perform a certain task, it makes a request to the receiver. When the receiver has finished executing the task it make an acknowledge to the sender that the task has been completed.

At this point the sender may initiate the next communication cycle. Since we have no global clock doing the job for us, we must introduce other means of synchronizing actions. We need a to have two parts in the circuit to represent the sender and the receiver. For the receiver we use a D latch that was already introduced earlier and for the sender we use a SR latch.

We already built a SR latch while building a D latch, as shown in Figure 12. A SR latch that is based on reversible optical architecture using direct logic elements, is depicted in Figure 17.

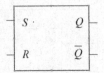

Fig. 17. Energy Efficient Optical SR Latch

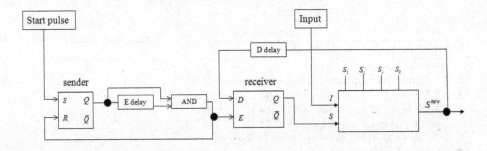

Fig. 18. 2-State Energy Efficient Optical Automata with Asynchronous Handshake Protocol

We build a asynchronous communication circuit using the SR latch as shown in Figure 18.

We use the E delay to ensure that the E input of the D latch is set to 1 only after the the automaton operated and generated a result S^{new}. We use the D delay to ensure that the D input of the D latch is set to S^{new} only once while E is 1, so S will not be updated in a loop to more than one new state S^{new}. It is common to use delays is asynchronous circuits [9,13], and it can be achieved optically for example by using mirrors or polarizers. After building the circuit there is no light at all, hence all the the inputs of all Fredkin gates are 0 and all other inputs of the circuit are 0. The first stage will be preparing the automaton for operation by initialize all Fredkin gates and set S_i, S_j, S_l, S_k. The automaton is ready to operate where in the initial state of the circuit the start state, input and the start pulse are all 0. The circuit communication cycle between the sender and receiver is described in Algorithm 2.

We created an energy efficient optical automata that can be used with asynchronous handshake protocol.

Cascading. In Figure 19 we can see two automata. The pulse to start the bottom automaton operation is when the R input of the SR latch of the upper automaton is set to 1. At this point the upper automaton already generated S^{new}, so it can be used as an input to the bottom automaton. The cascading is asynchronously

---

Algorithm 2. Communication cycle

1: Input I is ready.
2: Pulse is started.
3: The automaton operates and generates a result S^{new}.
4: The Q output of the SR latch is 1.
5: Pulse is stopped.
6: S^{new} is delayed; The 1 output of the SR latch is delayed.
7: S^{new} passes the delay and set to be the D input of the D latch.
8: The 1 output of the S-R latch passes the delay.
9: The E input of the D latch is set to 1.
10: R input of the SR latch is set to 1.
11: The Q output of the D latch is S^{new}.
12: The Q output of the SR latch is 0.
13: S is set to be the Q output of the D latch.
14: The E input of the D latch is set to 0.
15: The automaton is ready for the next input.

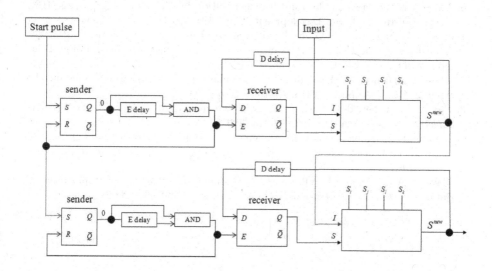

Fig. 19. Cascading automata circuit

and can be expanded to a chain of automata where one automaton's output is an input of the other.

All the presented are built using Directed Logic based on Fredkin gates, that known to be reversible, energy efficient and can be optically implemented.

References

1. Anter, A., Dolev, S.: Optical solution for hard on average #P-complete instances (using exponential space for solving instances of the permanent). Natural Computing 9(4), 891–902 (2010)
2. Bennett, C.H.: Logical reversibility of computation. Journal of Research and Development 17, 525–532 (1973)
3. Bennett, C.H.: The Thermodynamics of computation - A Review. International Journal of Theoretical Physics 21(3-4), 219–253 (1982)
4. Caulfield, H.J.: Zero-Energy Optical Logic: Can It Be Practical? In: Dolev, S., Oltean, M. (eds.) OSC 2009. LNCS, vol. 5882, pp. 30–36. Springer, Heidelberg (2009)
5. Caulfield, H.J.: Optics Goes Where No Electronics Can Go: Zero-Energy-Dissipation Logic. In: Dolev, S., Haist, T., Oltean, M. (eds.) OSC 2008. LNCS, vol. 5172, pp. 1–8. Springer, Heidelberg (2008)
6. Fitoussi, H.: Optical solutions for bounded NP problems, Master thesis, Ben-Gurion University of the Negev (2007)
7. Fredkin, E., Toffoli, T.: Conservative logic. International Journal of Theoretical Physics 21, 219–253 (1982)
8. Hardly, J., Shamir, J.: Optics insired logic architecture. Optics Express 15, 150–165 (2007)
9. Jacobson, H.: Asynchronous circuit design - A case study of framework called ack. Master thesis, Lulea University of Technology (1996)
10. Landauer, R.: Irreversibility and heat generation in the computing process. IBM Journal of Research and Development 5, 183–191 (1961)
11. Lee, J., Huang, X., Zhu, Q.: Embedding Simple Reversed-Twin Elements into Self-Timed Reversible Cellular Automata. Journal of Convergence Information Technology 6(1), 49–54 (2011)
12. Shamir, J., Caulfield, H.J., Micelli, W., Seymour, R.J.: Optical Computing and the Fredkin Gates. Applied Optics 25, 1604–1607 (1986)
13. Spars, J., Furber, S.: Principles of asynchronous circuit design - A systems perspective. Kluwer Academic Publishers (2001)
14. Thapliyal, H., Ranganathan, N.: Design of reversible sequential circuits optimizing quantum cost, delay, and garbage outputs. ACM Journal on Emerging Technologies in Computing Systems 6(4), Art. 14 (2010)
15. Zavalin, A.I., Shamir, J., Vikram, C.S., Caulfield, H.J.: Achieving stabilization in interferometric logic operations. Applied Optics 45, 360–365 (2006)

Object Signature Acquisition through Compressive Scanning

Jonathan I. Tamir[1], Dan E. Tamir[2], and Shlomi Dolev[3]

[1] Department of Electrical and Computer Engineering,
University of Texas at Austin, Austin, Texas 78712, USA
jtamir@utexas.edu
[2] Department of Computer Science,
Texas State University, San Marcos, Texas 78666, USA
dt19@txstate.edu
[3] Department of Computer Science, Ben-Gurion University of the Negev,
P.O. Box 653, Beer-Sheva 84105, Israel
dolev@cs.bgu.ac.il

Abstract. In this paper we explore the utility of compressive sensing for object signature generation in the optical domain. We use laser scanning in the data acquisition stage to obtain a small (sub-Nyquist) number of points of an object's boundary. This can be used to construct the signature, thereby enabling object identification, reconstruction, and, image data compression. We refer to this framework as *compressive scanning* of objects' signatures. The main contributions of the paper are the following: 1) we use this framework to replace parts of the digital processing with optical processing, 2) the use of compressive scanning reduces laser data obtained and maintains high reconstruction accuracy, and 3) we show that using compressive sensing can lead to a reduction in the amount of stored data without significantly affecting the utility of this data for image recognition and image compression.

Keywords: Digital Signal Processing, Optical Signal Processing, Compressive Sensing, Shape Representation, Object Signature, Optical SuperComputing.

1   Introduction

Technology for 2D and 3D laser scanning is currently utilized for acquiring object information and representing its surface. Often, this information is used as a part of geometry-based graphics, where points on the object boundary, obtained throughout the scan, are considered as vertices in a polygonal mesh model of the object [1]. Alternatively, the points obtained can be used for numerous shape representation methods, including those derived from object signature representation [2]. Compressive sensing (CS) of an object's signature has been reported in [3], [4]. Nevertheless, CS of signatures is currently done using digital processing techniques applied to an image of the object. In this paper we explore the utility of moving CS to the optical domain. In the data acquisition stage, i.e., the

S. Dolev and M. Oltean (Eds.): OSC 2012, LNCS 7715, pp. 105–116, 2013.

laser scan, we obtain a small number of samples of the object's boundary. This scan can be used for accurate reconstruction of the signature, thereby enabling object reconstruction, image data compression, and object recognition.

Compressed sensing approaches based on instant 2D capturing of spatial images in the optical domain via compressed transformations are reported in [5]. Nevertheless, our work differs from the work reported in [5] due to the fact that our method captures object signatures rather than object's raster image. Hence, we avoid the costly and error-prone process of segmentation, object identification, and contour following that precede traditional signature acquisition in the raster domain [3].

The contributions of the paper are the following:

1. CS is applied at the laser scanning stage to construct an object signature, thereby replacing parts of the digital processing with efficient optical processing.
2. The use of CS enables a reduction in the amount of laser data obtained while maintaining high reconstruction accuracy.
3. We show that using CS can lead to a reduction in the amount of stored data without significantly affecting the utility of this data for image recognition and image compression.

As far as we know, and based on an extensive literature review, this is the first use of CS in the context of laser scan for object signature acquisition.

In this paper we explore the utility of using optical processing as a part of acquiring an object signature using CS methodology. This is enabled via random sampling of an object's boundary at the laser acquisition stage. Furthermore, we analyze the reconstruction and identification performance of this approach by comparing the resultant distortion to that of other methods, such as interpolation and thresholding.

The remainder of this report is organized as follows. Section 2 describes shape representation methods, concentrating on object signature construction in the image and optical domains. In Section 3, we provide an overview of CS theory, and apply it to object signature reconstruction in Section 4. The experimental setup is described in Section 5. Section 6 explores the identification capability of CS-based signatures and compares its distortion to that of interpolation and discrete cosine transform (DCT)-based compression. Finally, Section 7 gives concluding remarks and a roadmap for future work.

2   Object Signature

One of the active areas of research in image processing is shape representation [2], [6], [7], [8]. Generally, the procedure for shape representation includes identifying objects, marking the pixels in the boundary of objects, and using compact and efficient methods for representing these pixels. Of interest are shape representation techniques that are invariant to affine transformations, i.e. combinations of translation, rotation, and scaling [6]. Some of the commonly used methods for shape representation include chain codes, Fourier descriptors, B-splines, and moments [6], [2]. The signature of object is a set of distances from a reference point – generally the object centroid – to its boundary [2], [7]. Baggs and Tamir as well as Keogh present several variants of object signature [7], [8]. This representation is invariant to affine transformations. In addition, Baggs and Tamir demonstrate methods for dealing with non-linear warping of objects [7]. Furthermore, the representations proposed can be efficiently used for object recognition, alignment/registration, and compression [7]. An interesting question is the resilience of these methods to compression applied to the object signature. That is, only a compressed version of the signature is stored. This has been addressed in [3] and is further explored in the current paper.

2.1 Shape Representation Using Object Signature

In general, the term object signature relates to a set of measurements of the distances from a fixed point to object boundary elements. The term object-signature, however, is overloaded. In a fixed angle signature, the distances from the object centroid to its boundary are measured in increments of equal angles. A problem with the fixed angle method occurs with certain concave objects where the same angle might yield more than one distance to points on the contour. In practice, the *equal-angle* approach does not require convexity and can be applied to certain star-shaped objects [9]. A second approach is to measure the distance from the centroid to every pixel on the contour of the object. The *every-pixel* method removes the requirement for star-shaped convexity and can use an arbitrary centroid with arbitrary objects. On the other hand, the every-pixel approach generates a variable number of samples which depends on the object and is sensitive to scaling. This might necessitate non-linear warping, referred to as dynamic space warping, of the signature [7], [8]. This work uses the equal-angle method for signature generation. Figure 1 shows two objects and their corresponding signatures in space and in frequency (DCT).

Object representation via its signature has several advantages over other shape representation methods with respect to object matching, reconstruction, and compression. The main advantage is that it is less sensitive to noise. This is due to the fact that a small change in shape, which may be due to noise, causes minimal change to the distance between the centroid and contour points. In addition, this representation converts a two-dimensional signal into a one-dimensional signal with minimal loss of information and provides a computationally efficient framework.

2.2 Acquiring an Object Signature in the Image Domain

In order to obtain an object signature from an image, the image has to be segmented, objects have to be identified, and the pixels that reside on object contours have to be marked [6]. Image segmentation can be implemented in several different ways [2]. Following the segmentation, connected component labeling and contour following algorithms are applied to identify boundary pixel

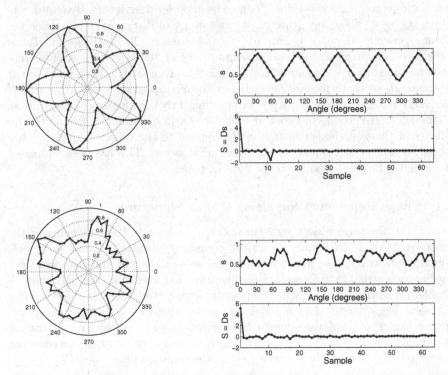

Fig. 1. Objects and their corresponding signatures in space and frequency (DCT)

and construct the object signature [6]. One inherent problem with this approach is that it depends on the accuracy of the segmentation. But, many segmentation algorithms introduce deficiencies such as sensitivity to gray level changes, issues related to the introduction of artificial edges by edge detection procedures, and performance degradation due to noise. In general, the problem of image segmentation is an open image processing problem and the incompleteness of current segmentation technology affects subsequent image processing procedures. Researchers have observed that morphological operations, edge detection, and segmentation techniques might yield better results using laser scan data [10], [11]. Practically, when object features are related to sharp changes in elevation rather than changes in gray level, segmentation and edge detection procedures are more suited to the laser measurement domain. To this end, in the next section we introduce the topic of acquiring objects' signatures using laser scanning. Furthermore, in Section 4 we apply CS to the process of obtaining the signature in the laser measurement domain.

2.3   Acquiring an Object Signature Using Laser Scanning

In this work we consider *pseudo*-2D objects. A *pseudo*-2D object is a 3D object where one of its dimensions (say Z) has a relatively low variance for a large contiguous set of points in the object surface. For example, consider a tank in the desert that is scanned by an airplane-mounted laser. There is a large difference between the values of Z-coordinates on the tank's top-surface and the Z-coordinates of the desert background. In this case the results of 3D laser scanning can be accurately used to segment the object and identify its endpoints. This capability is often built into the acquisition hardware [10].

In this paper we simulate a laser device that provides only the endpoints of objects within a laser scan (that is, points with a sharp change in the Z-coordinate) [10]. For example, one may use a line scan camera along with a laser line generator (LLG) such as the 13LT family of LLGs described in [11]. Furthermore, we assume that the laser scan is implemented in a way that is similar to scan conversion, as shown in Figure 2. The laser emits horizontal and/or vertical lines and the endpoints of the object within the lines are recorded.

In general, using laser scan rather than a single full-image capture increases the acquisition time. Nevertheless, for shape representation via signature, our method has the advantage of capturing the signature in the optical domain, saving the time associated with preprocessing operations such as segmentation, connected component labeling, and contour following [3]. Moreover, in many cases, (e.g., identification of *pseudo*-2D objects via aerial photography) laser scan (or a hybrid of laser scan and image capture) can increase the accuracy of the above mentioned preprocessing operations.

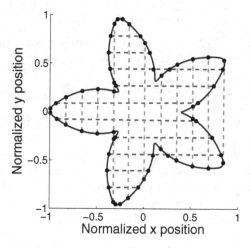

Fig. 2. Horizontal and vertical laser scanning of an object

3   Compressive Sensing Theory

Compressive sensing is an alternative to the traditional Shannon-Nyquist based analog to digital conversion (ADC) [12], [13], [14]. Under this approach, rather than sampling an analog signal at its Nyquist rate, a small number of measurements or linear combinations of its samples are obtained and quantized. This sampling method has an advantage over the traditional sampling provided that it results in sufficient reconstruction accuracy and the number of samples is smaller than the number of samples required by the Nyquist Theorem. Candes and Donoho, as well as many other researchers, have investigated numerous aspects of CS and their applications in the fields of sensing, imaging, and signal processing [15], [16]. In this section we examine CS theory; that is, we examine the requirements posed on a signal and the number of measurements required in order to obtain reconstruction with sufficient quality [12], [13].

Consider an n-dimensional vector (signal) $\mathbf{s} \in \mathbb{R}^n$. The goal is to obtain a representation of $\mathbf{s}$ in the form of $\mathbf{x} \in \mathbb{R}^m$ such that: 1) $\mathbf{x}$ is a set of linear measurements of $\mathbf{s}$, 2) $\mathbf{s}$ can be reconstructed from $\mathbf{x}$, and 3) $m \ll n$. Hence, the vector $\mathbf{x}$ is of the form

$$\mathbf{x} = A\mathbf{s}. \tag{1}$$

That is, $\mathbf{x}$ is the result of correlating $\mathbf{s}$ with the waveforms A_i, $i \in [1 : m]$. In other words, $\mathbf{x}$, the representation of $\mathbf{s}$, is obtained via sensing of $\mathbf{s}$ using m vectors of the form $A_i \in \mathbb{R}^n$. CS theory states the conditions on $\mathbf{s}$, $\mathbf{x}$, m, and A such that $\mathbf{s}$ can be reconstructed from $\mathbf{x}$ with acceptable quality. Next, we discuss these conditions. Let $\|\cdot\|_p$ denote the l_p-norm of $\mathbf{a}$. Also, let $|\mathbf{a}|$ denote the cardinality of $\mathbf{a}$. A signal $\mathbf{s}$ is k-sparse if the cardinality of its support complies with

$$|\{i : s_i \neq 0\}| \leq k. \tag{2}$$

According to Candes $et\ al$ [17], one could almost surely recover the signal $\mathbf{s}$ from the set of measurements $\mathbf{x}$ by solving the convex program

$$\underset{\tilde{\mathbf{s}} \in \mathbb{R}^n}{\mathrm{argmin}} \quad \|\tilde{\mathbf{s}}\|_1 \quad \text{s.t.} \quad \mathbf{x} = A\tilde{\mathbf{s}} \tag{3}$$

provided that A satisfies the Restricted Isometry Property (RIP) [17] and that

$$m \geq Ck\log(n), \tag{4}$$

where C is a constant. The theoretical value of C and its bounds have been investigated and documented in several research papers [12], [13]. In practice, often the value $C = 4$ gives good signal to noise ratio (SNR) results [12].

Using traditional sampling theory, n, the number of samples required for exact reconstruction of $\mathbf{s}$, must comply with the Nyquist theorem. The novelty of CS is that the number of required measurements m is much smaller than the number of required samples n. Additional theoretical results and practical implementations of CS theory can be found in [12], [13].

4   CS Signature

Let $\mathbf{s} = \begin{bmatrix} s_1\ s_2 \cdots s_n \end{bmatrix}^T$ denote the rotation-dependent object signature, described in Section 2.1, with sampling rate $\frac{n}{360}$ samples per degree. By definition,

$0 < s_i \leq 1$ for all $i \in [1 : n]$, and $\mathbf{s}$ is not sparse. By the incoherency of time and frequency [17], we expect the signature's frequency-domain representation to be sparse. Indeed, the DCT $\mathbf{S}$ of $\mathbf{s}$ is both sparse and rapidly decaying, with most energy concentrated at low frequencies. Let $D \in \mathbb{R}^{n \times n}$ denote the DCT matrix. Then

$$\mathbf{S} = D\mathbf{s}. \tag{5}$$

Random rows of the DCT matrix satisfy RIP [16]. Let $A \in \mathbb{R}^{m \times n}$ represent m randomly chosen rows of D^T – the inverse DCT matrix – and let

$$\mathbf{y} = A\mathbf{S}. \tag{6}$$

If m satisfies (4), then $\mathbf{S}$ can be approximately recovered by

$$\underset{\tilde{\mathbf{S}} \in \mathbb{R}^n}{\operatorname{argmin}} \quad \|\tilde{\mathbf{S}}\|_1 \quad \text{s.t.} \quad \|\mathbf{y} - A\tilde{\mathbf{S}}\|_2 \leq \epsilon. \tag{7}$$

Note that (6) is equivalent to choosing m random points of $\mathbf{s}$. Thus, given random samples of the object signature $\mathbf{s}$, we can recover the rotation-invariant DCT $\mathbf{S}$ through (7). This can be used for object recognition by comparing the recovered signal to a database of DCT signals, or it can be used to reconstruct an object up to a rotation and scaling ambiguity.

Given an object with centroid (x_c, y_c) and random angles,

$$\theta_i \sim U[0, 2\pi], \quad i \in \left[1 : \frac{m}{2}\right], \tag{8}$$

we can obtain m random samples of its signature $\mathbf{s}$ by scanning through (x_c, y_c) at each angle θ_i. From these angles, we build the sensing matrix A and use CS techniques to recover $\mathbf{S}$.

5   Experimental Setup

We constructed a library of star-shaped objects, where the object centroid lies in the star-shaped kernel [9] and is not co-linear with any object vertex, by generating discrete signals with n points, representing uniformly sampled signatures with angular sampling rate $f_s = \frac{n}{360}$ samples per degree. Here, f_s is the highest Nyquist rate among all objects in the library. The library includes basic shapes in addition to objects with randomly generated contours.

To simulate laser acquisition of an object, we obtain a set of boundary points $\mathcal{P}$ through horizontal and vertical scanning with scan separation Δy and Δx, respectively. The object's centroid is estimated as (x_c', y_c') using these measurements. Note that the distribution of points acquired is a function of both the scan separation and the object itself. Thus, the points are unlikely to be equally spaced along the object perimeter.

Our library contained 10 object signatures (5 basic shapes and 5 random shapes) with $n = 64$ (approximate Nyquist-rate sampling). For this work, we compared three methods for acquiring and reconstructing the DCT signature of the objects in the library and compared the distortion for varying values of $m = |\mathcal{P}|$. Distortion was measured as the mean-squared error (MSE) of the reconstructed signature and the closest-matching signature in the library. Equivalently, we report the effective signal to noise ratio (SNR), defined as

$$\text{SNR [dB]} = 20 \log_{10} \frac{\|\mathbf{s}\|_2}{\|\mathbf{s} - \tilde{\mathbf{s}}\|_2}, \tag{9}$$

where $\tilde{\mathbf{s}}$ is the reconstruction of $\mathbf{s}$. The SNR can be used as a measure of similarty between the signatures $\tilde{\mathbf{s}}$ and $\mathbf{s}$. Note that if $\tilde{\mathbf{s}} = \mathbf{s}$, then the SNR is infinite. In addition, the SNR can be used to measure reconstruction error when reconstructing the object from $\tilde{\mathbf{s}}$.

5.1   Experiment 1

For the first experiment, we performed an additional scanning stage, in which the laser scans through the estimated centroid (x'_c, y'_c) at angles θ_i, $i \in \left[1 : \frac{m}{2}\right]$. These m points are converted to polar coordinates with respect to the center. There are two cases to consider for the second scanning stage.

Case 1. Uniform sampling: In the first case, m equally spaced angles are chosen, i.e. $\theta_i = (i - 1)\frac{n}{m}$ for all i. These points are then resampled to n points and normalized. For $m < n$, this represents sub-Nyquist sampling, and we expect aliasing. We refer to this case as reconstruction via resampling, or interpolation.

Case 2. Random sampling: In the second case, m angles are chosen according to (8). Using these points, we perform CS recovery as described in Section 4. We refer to this case as reconstruction via two-stage CS.

5.2   Experiment 2

Due to the variation in object perimeters, we can consider the points in $\mathcal{P}$ as random samples of the object boundary. Thus, rather than performing a second scanning stage through the object's centroid, we converted the points in $\mathcal{P}$ to polar coordinates with (x'_c, y'_c) as the origin to build the sensing matrix A. We then approximately recovered the object signature in the DCT domain through CS, as described by (7). We refer to this approach as single-stage CS.

5.3   Experiment 3

As a baseline comparison, we took the DCT $\mathbf{S}$, retained the k largest values, and replaced the remaining $n - k$ samples with zeros. We then compared distortion to the original signature. As a rule of thumb, we expect similar distortion to the CS case for $m \approx 4k$ [13]. We refer to this approach as reconstruction via thresholding.

The full process of acquisition and reconstruction is described in Figure 3. With this acquisition framework, we are able to compare the performance of CS recovery to standard resampling, i.e. interpolation and decimation, under different choices of parameters.

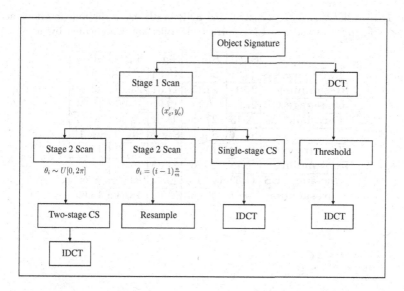

Fig. 3. Acquisition and reconstruction of object signature

6   Results

Table 1 shows the recovery SNR across all objects (averaged over 500 trials) for $m = 48$ and $k = 12$, and indicates whether the correct object was identified (i.e., the signature corresponding to the true object had the highest SNR). As seen, all methods except single-stage CS correctly identified every object in the library. For nearly every object, the SNR among the two cases in Section 5.1 are in excess of 15 dB. There is a 2-10 dB drop between the basic shapes and the random shapes, indicating that additional structure (e.g. redundancy) may be exploited in the former group. The two-stage CS recovery affords an additional 2-5 dB gain for the basic objects in comparison to resampling. Figure 4 shows the accurate reconstruction performance across all methods for the two objects shown in Figure 1.

Because the single-stage CS approach does not use random samples, there is potential loss in incoherency of the sensing matrix A. Thus, we must use more samples to achieve an equivalent SNR. Nonetheless, the single-stage approach saves considerably in acquisition time, and it does not require *a priori* knowledge of the object's centroid. Thus it may be beneficial to reduce the scan separation to acquire more samples while skipping the second scanning stage.

As expected, the achieved SNR from acquiring the entire signature and thresholding the DCT to $k = \frac{m}{4} = 12$ samples matches or exceeds the SNR of all other methods. Although this approach provides the optimal lossy compression among the proposed methods, it requires full knowledge of the signature.

Table 1. Identification and reconstruction SNR in dB using the four methods across all objects for $m = 48$ and $k = 12$. Incorrect identification is indicated by an ×.

	Sig 1	Sig 2	Sig 3	Sig 4	Sig 5
Two-stage CS	30.31	26.50	18.64	25.36	25.62
Resampling	27.05	22.15	17.81	23.29	22.89
One-stage CS	27.59	25.11	19.01	17.06	16.83
Thresholding	29.13	27.86	25.91	28.50	30.64
	Sig 6	Sig 7	Sig 8	Sig 9	Sig 10
Two-stage CS	17.60	19.95	20.78	16.20	14.35
Resampling	16.78	20.06	21.35	16.51	12.58
One-stage CS	11.89	14.73 ×	13.82	9.09 ×	8.39
Thresholding	20.87	26.36	21.68	18.21	16.69

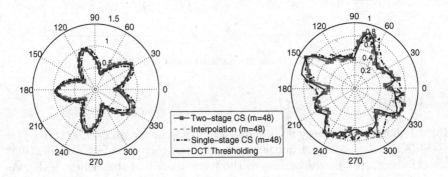

Fig. 4. Reconstruction ($n = 64$) of a basic object and random object using the four described methods ($m = 48$, $k = 12$)

Figure 5 shows the achieved SNR of the two cases described in Section 5.1. The objects depicted in Figure 1 were reconstructed for different choices of m (averaged over 500 trials). At $m = n = 64$, both approaches accurately reconstruct the objects with negligible MSE. The figure shows the clear gap in SNR between the two types of objects. Furthermore, it highlights the SNR gain when reconstructing via CS. Note that our experiments did not incorporate noise in the laser scan acquisition. In a more realistic scenario, the acquired signature is a noisy version of the real signature. Hence, we expect degradation of reconstruction and identification accuracy in this setting.

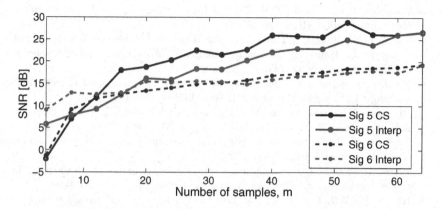

Fig. 5. Two-stage CS and Resampling recovery SNR in dB over different values of m for two object signatures

7 Conclusions and Future Work

We have introduced a new and innovative framework for *compressive scanning* of objects' signatures in the optical domain. We used the DCT as a sparse representation domain and performed several experiments comparing CS reconstruction, interpolation, and DCT-based compression to show the viability of the new approach. Furthermore, the methodology presented enables a reduction in the amount of laser data obtained and stored while maintaining high reconstruction accuracy. In the future, we plan to investigate the utility of other sparse representations e.g., total variation norm, wavelet transform, and the Fourier transform. In addition, we plan to investigate the use of the cyclic auto correlation of the signature in order to resolve rotation and scaling on the performance of compressive scanning in the context of image recognition and compression. Our analysis assumed no noise in the signature acquisition process. We plan to assess the impact of noisy laser-scan measurements on the accuracy of signature based identification and reconstruction. Another subject planned for further research is dynamic adjustment of the sampling rate up to the point of diminishing returns. Under this approach, the initial number of samples taken from an inspected object is small. Then, the object is compared to a library of objects at increasing levels of resolution (to the point of dimished return) and the resolution that yields the highest matching score between the scanned object and any library element is used for the CS resolution. Finally, we plan to explore the use of variable-length signatures, which is analogous to the every-pixel approach for digital images.

References

1. Frueh, C., Zakhor, A.: Constructing 3D city models by merging ground-based and airborne views. In: Proceedings of the 2003 IEEE Computer Society Conference on Computer Vision and Pattern Recognition, vol. 2, pp. II-562–II-569 (June 2003)

2. Gonzalez, R.C., Woods, R.E.: Digital Image Processing, 3rd edn. Prentice-Hall, Inc., Upper Saddle River (2006)
3. Tamir, D.E., Shaked, N.T., Geerts, W.J., Dolev, S.: Compressive sensing of object-signature. In: Dolev, S., Oltean, M. (eds.) OSC 2010. LNCS, vol. 6748, pp. 63–77. Springer, Heidelberg (2011)
4. Ye, J.C.: Compressed sensing shape estimation of star-shaped objects in fourier imaging. IEEE Signal Processing Letters 14(10), 750–753 (2007)
5. Rivenson, Y., Stern, A., Javidi, B.: Compressive fresnel holography. Display Technology. Journal of Display Technology 6(10), 506–509 (2010)
6. Pavlidis, T.: Algorithms for graphics and image processing. Digital system design series. Computer Science Press (1982)
7. Baggs, R., Tamir, D.E.: Image registration using dynamic space warping. In: Artificial Intelligence and Pattern Recognition 2008, pp. 128–135 (2008)
8. Keogh, E., Wei, L., Xi, X., Hee Lee, S., Vlachos, M.: LB Keogh supports exact indexing of shapes under rotation invariance with arbitrary representations and distance measures. In: VLDB, pp. 882–893 (2006)
9. Arkin, E.M., Chiang, Y.J., Held, M., Mitchell, J.S.B., Sacristan, V., Skiena, S.S., Yang, T.C.: On minimum-area hulls (1998)
10. Hug, C.: Extracting artificial surface objects from airborne laser scanner data. In: Automatic Extraction of Man-Made Objects from Aerial and Space Images II, pp. 203–212 (1997)
11. Kirchhoff, S.: Laser diode collimator flatbeam (2011), http://www.sukhamburg.com/onTEAM/pdf/cam_cat_44-45_en.pdf
12. Donoho, D.: Compressed sensing. IEEE Transactions on Information Theory 52(4), 1289–1306 (2006)
13. Candes, E., Tao, T.: Near-optimal signal recovery from random projections: Universal encoding strategies. IEEE Transactions on Information Theory 52(12), 5406–5425 (2006)
14. Porat, B.: A course in digital signal processing. John Wiley (1997)
15. Lustig, M., Donoho, D.L., Santos, J.M., Pauly, J.M.: Compressed sensing MRI. IEEE Signal Processing Magazine (2007)
16. Elad, M.: Optimized projections for compressed sensing. IEEE Transactions on Signal Processing 55(12), 5695–5702 (2007)
17. Candes, E., Wakin, M.: An introduction to compressive sampling. IEEE Signal Processing Magazine 25(2), 21–30 (2008)

All-Optical SOA-Assisted 40 Gbit/s DQPSK-to-OOK Format Conversion

Mirco Scaffardi[1], Valeria Vercesi[1], Sergio Pinna[2], and Antonella Bogoni[1]

[1] CNIT, Pisa, Italy
mirco.scaffardi@cnit.it
[2] TeCIP, Scuola Superiore Sant'Anna, Pisa, Italy

Abstract. An SOA-assisted 40 Gbit/s DQPSK-to-OOK format converter is presented. An SOA-based amplification stage, after conversion, provides negative penalty (>1.5 dB) and extinction ratio improvement (up to 6 dB), also after propagation, making the converted OOK-signals suitable for optical processing.

Keywords: Format Conversion, Nonlinear Optical Processing, Semiconductor Optical Amplifiers.

1 Introduction

Telecommunications optical networks are characterized by different requirements depending on the geographical reaches required. The wide and metro optical networks are characterized by systems with fast reconfigurability, and high capacity and by modulation formats that are more robust to the detrimental consequences of the nonlinear effects experienced by the signal during the propagation. For instance, differential quaternary phase-shift-keying (DQPSK) is widely investigated for transmission on long haul and at high bit rates [1]. On the other side, local-area networks require large connectivity and bandwidth flexibility at much lower cost. Therefore simple and cost-efficient transponders delivering on-off keying (OOK) format is the most economic way to deploy connectivity on a large scale as the case of wavelength division multiplexing passive optical network (WDM-PON) [2].

Nowadays, to bridge optical traffic from the wide and metro to the access networks, expensive and power hungry opaque edge nodes are used consisting of high speed front-end optical- electro-optical (OEO) transceivers, and digital signal processors. In this scenario there is a clear need for new optical edge nodes that transparently interconnect the different networks avoiding expensive and power consuming opaque edge nodes.

With reference to Fig. 1, an all-optical solution to adapt the wide/metro traffic to the access traffic, has to provide the format, bit rate and wavelength conversion in the optical domain. In the case of incoming DQPSK coded signals, the format conversion can be carried out by means of conventional DQPSK demodulators that are passive integrated optical circuits providing the corresponding OOK data streams.

S. Dolev and M. Oltean (Eds.): OSC 2012, LNCS 7715, pp. 117–122, 2013.

Then, already demonstrated optical wavelength conversion schemes [3] can adapt the wavelength of each OOK channel to the access network requirements, by using a battery of (tunable) continuous wave (CW) lasers. If the converted OOK signals have a bit rate higher than the access traffic bit rate, the bit rate conversion can be obtained in the wavelength converter by using optical pulsed clocks at repetition rate equal to the desired final bit rate, instead of CW beams.

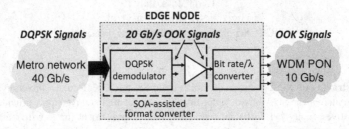

Fig. 1. Architecture of a DQPSK-to-OOK optical edge node based on the SOA-assisted format converter

In order to provide the desired signal power levels to the bit rate/wavelength converter, the format conversion block requires a final amplification stage. Typically, at the output of the DQPSK demodulator the extinction ratio (ER) of the OOK signals is reduced due to a not negligible floor on the low level. Normally it is compensated by a balanced photodetection, but in this case the signals are kept in the optical domain and they interact with the additional noise introduced by the optical amplifier significantly impacting on the penalty introduced by the edge node and consequently on the quality of the traffic provided to the access networks.

Here we demonstrate that the use of one or two cascaded semiconductor optical amplifiers (SOAs) for implementing the amplifying stage, allows to simultaneously exploit the SOA gain for the amplification and the SOA nonlinearities for improving the OOK signal extinction ratio, resulting in a negative penalty of the amplification stage and consequently in a penalty reduction of the format converter.

The paper is organized as follow. In section 2, we describe the proposed SOA-assisted format converter working principle and the experimental setup used to verify the effectiveness of the system in the case of 40 Gb/s DQPSK and 20 Gb/s OOK input and output signals respectively, both with and without propagation of the generated OOK frames in optical fiber and for input signal-to-noise ratio in the range 30-40 dB. In section 3 the experimental results are shown and in the last section conclusions are reported.

2   Working Principle and Experimental Setup

The architecture of the SOA-assisted format converter is shown in Fig. 2. An input DQPSK signal is converted to an OOK signal with a format converter, i.e. a DQPSK demodulator. To make the signal suitable for all-optical processing after demodulation, an amplification stage based on SOA is used. The amplification stage includes a band pass filter (BPF) and a polarizer (POL).

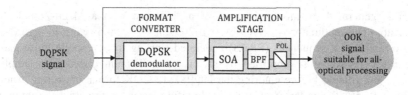

Fig. 2. SOA-assisted format converter architecture

The floor on the zero level of the demodulated OOK signal is reduced exploiting self-phase modulation (SPM)-induced polarization rotation and filtering [4]. When the signal enters the SOA it experiences spectral shift depending on the signal power. The zero level experiences a shift lower than the one level. Therefore, by adjusting the BPF central wavelength it is possible to reduce the zero level power. The input signal experiences also power dependent self polarization rotation. Therefore, by properly adjusting the orientation of the polarizer, it is possible to cut the power on the zero level. Both SPM and polarization rotation allows improving the signal extinction ratio. Moreover the SOA-gain saturation can improve also the equalization of the OOK signal one level [5].

The experimental setup implemented to investigate the penalty introduce by the SOA stage on the OOK demodulated signal is shown in Fig. 3. The 40 Gbit/s DQPSK

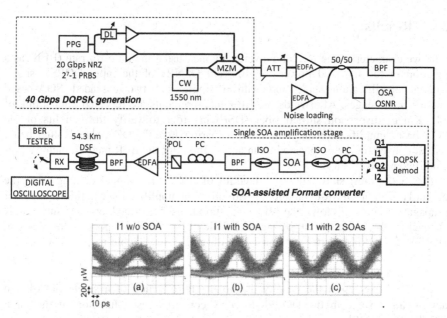

Fig. 3. Top: Experimental setup. Bottom: Eye diagrams of the converted OOK signals without amplification (a), with a single (b) and two (c) cascaded amplification stages after propagation through 54.3 Km DSF. (PPG: pulse pattern generator; DL: delay line; MZM Mach Zehnder modulator; ATT: attenuator; BPF: band pass filter; OSA: optical spectrum analyzer; PC: polarization controller; ISO: optical isolator; POL: polarizer; RX: pre-amplified receiver).

signal is generated by modulating a CW at 1550 nm with a Mach-Zehnder I-Q modulator. The modulator is driven by two 20 Gbit/s 2^7-1 non-return-to-zero (NRZ) pseudo-random bit sequences (PRBS) generated starting from a single pulse pattern generator. The signal is amplified by means of an Erbium doped fiber amplifier (EDFA) and filtered before entering the DQPSK-to-OOK format converter. Another EDFA and a 50/50 coupler allow noise loading for measurement with different optical signal-to-noise ratios (OSNRs) of the input DQPSK signal. The format converter is composed by an integrated 40 Gb/s DQPSK demodulator provided by Optoplex, whose working point is determined by thermal control, followed by the SOA-based amplifying stage. Each single SOA amplification stage includes a polarization controller at the SOA input for optimizing the signal polarization state, optical isolators at the SOA input and output for counteracting optical reflections and a band-pass filter (BW: 1 nm), a polarization controller and a polarizer at the SOA output. The filter has to be slightly detuned compared to the signal, in order to select the signal portion corresponding to the signal high level that experienced the biggest amount of SPM, simultaneously reducing the signal low level. SPM induces also signal polarization rotation, therefore the polarization controller and the polarizer can improve the zero level as well. In the experiment we used a Kamelian SOA with 6.7 dB noise figure, 10 dBm output saturation power and 24 dB gain.

3   Results

First we characterize the format converter by measuring the bit error rate (BER) as a function of the received power for different OSNRs of the input DQPSK signal without amplification stage and with a single and two cascaded SOA-based amplification stages. Afterwards the BER measurements are repeated after a link of 54.3 Km of dispersion shifted fiber (DSF) in order to verify the benefits of the SOA-based amplification stage in a real scenario where the OOK signals after being format converted, are propagated in an access link. For the BER measurements a pre-amplified receiver is used.

Fig. 3 (bottom) shows the eye-diagram of the OOK channel coming from the output I1 of the DQPSK demodulator with different amplification conditions and after propagation. Fig. 4 shows the BER measurements performed on the same OOK channel, with an input DQPSK signal OSNR of 34 dB. We report all the cases of DQPSK-to-OOK converter without and with single and double amplification stage, with and without propagation. We can note that the single amplification stage introduces a penalty reduction of about 1.5 dB.

In Fig. 5 we report the penalty due to the introduction of a single SOA-based amplification stage in the DQPSK-to-OOK converter, as a function of the input DQPSK signal OSNR.

The penalty is calculated at BER = 10^{-9}. The curve shows a negative penalty over a 10 dB OSNR range (from 30 dB to 40 dB) with an absolute value up to 1.6 dB, therefore the amplification process does not reduce the quality of the converted signal. On the contrary the SOA reduces the low level floor of the OOK signal resulting in an improvement of the DQPSK-to-OOK converter performances.

The introduction of two cascaded amplification stages does not significantly improve the penalty reduction of the DQPSK-to-OOK converter as shown in Fig. 4 both in the case with propagation and without. The impact of the second amplification stage is more relevant on the extinction ratio improvement of the OOK signals at the output of the DQPSK demodulator. In fact, as shown in Fig. 3 (right-c) the second SOA eliminates the low level floor and reduces the noise on the one-level. As reported in Fig. 6, ER of the input DQPSK signal from 4 dB to 7 dB can be increased beyond 8 dB up to 12 dB. This represents an improvement of more than 2 dB compared to the single amplification stage.

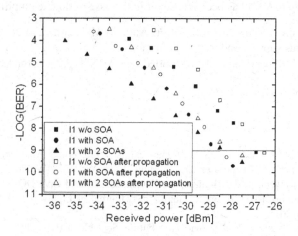

Fig. 4. BER vs. received power of the DPSK-to-OOK converted signal without amplification stage, with a single SOA-based amplification stage and with two cascaded SOA-based amplification stages after propagation through 54.3 Km DSF

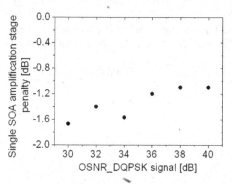

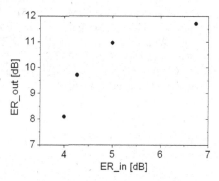

Fig. 5. Negative penalty due to the introduction of a single SOA-based amplification stage in the DQPSK-to-OOK converter

Fig. 6. Output extinction ratio vs. input extinction ratio when two cascaded SOAs are employed

4   Conclusions

An enhanced all-optical 40 Gbit/s DQPSK-to-OOK format converter is presented. It includes a SOA-based amplification stage with a negative penalty that results in an extinction ratio improvement of the converted signals making the format converter suitable for applications where optical signal processing follows the format conversion stage, as the case of all-optical edge nodes adapting the metro traffic to access networks.

Penalty reduction due to a single and double amplification stage higher than 1.5 dB has been obtained both without propagation and after 54.3 Km-long DSF. ER improvement up to 6 dB has been measured in the case of double amplification stage.

Acknowledgments. The authors thank dr. Antonio Malacarne and dr. Emma Lazzeri for their precious support and useful discussions.

This work has been partially supported by ARNO project funded by Tuscany Region-Italy.

References

1. Winzer, P.J., Raybon, G., Song, H., Adamiecki, A., Corteselli, S., Gnauck, A.H., Fishman, D.A., Doerr, C.R., Chandrasekhar, S., Buhl, L.L., Xia, T.J., Wellbrock, G., Lee, W., Basch, B., Kawanishi, T., Higuma, K., Painchaud, Y.: 100-Gb/s DQPSK Transmission: From Laboratory Experiments to Field Trials. Journal of Lightwave Technology 29(15), 2259–2266 (2011)
2. Kani, J.-I.: Enabling Technologies for Future Scalable and Flexible WDM-PON and WDM/TDM-PON Systems Optical Digital Signal Processing in a Single SOA Without Assist Probe Light. IEEE Journal of Selected Topics in Quantum Electronics 16(5), 1290–1475 (2010)
3. Chung, H.S., Inohara, R., Nishimura, K., Usami, M.: All-optical Multi-Wavelength Conversion of 10 Gbit/s NRZ/RZ Signals Based on SOA-MZI for WDM Multicasting. Electronics Letters 41(7) (March 2005)
4. Porzi, C., Scaffardi, M., Potì, L., Bogoni, A.: Optical Digital Signal Processing in a Single SOA Without Assist Probe Light. IEEE Journal of Selected Topics in Quantum Electronics 16(5), 1469–1475 (2010)
5. Contestabile, G.: All-Optical Signal Regeneration using SOAs. In: Proceedings ACP 2010, December 8-12, pp. 7–8 (2010)

Obituary
H. John Caulfield

Yaroslavsky L.

Dept. of Physical Electronics, School of Electrical Engineering, Tel Aviv University

On Febr. 3, 2012, a very sad message came: John Caulfield, one of pioneers, passionate enthusiasts and popularizers of holography and optical information processing, passed away.

We got personally acquainted some 30 years ago on, as far as I remember, a conference on holography in Varna, Bulgaria. At that time I was working in USSR Academy of Sciences and I was a so called "nevyezdnoy", which meant, in Russian, a person not allowed to travel outside the Soviet bloc. I was therefore prevented from participating in conferences and meetings in Western countries and for me meeting and getting acquainted with John was a breath of fresh air and freedom. Then we met whenever John attended conferences in the former Soviet Union or visited the USSR on other occasions. These meetings were always tremendous support, encouragement and stimulation for me.

Since then, we met many times on conferences and had been regularly communicating by e-mail up to the very end. He was always full of ideas. Whenever and whatever I told him he always immediately exclaimed: "Great!", "Wonderful!". This was his way to communicate with people: very friendly, encouraging, collaborating. And almost always he advanced new ideas and suggested writing a paper. He had a very broad horizon of interests. His physical intuition was unbelievable. And indeed, two of our short meetings on conferences resulted in several journal and conference papers I am proud of. Many of his friends can tell similar stories.

John was a great personality. He wrote and edited many books, founded scientific journals. He had a profound influence on the field of holography and informational optics. We will miss his powerful presence.

Dr. H. John Caulfield
Outstanding Scientist and Magnificent Friend

Dr. R. Barry Johnson

Alabama A&M University
Normal, Alabama
22 October 2012

After working in industry for many years, Dr. H. John Caulfield and I founded the Center for Applied Optics (CAO) at The University of Alabama in Huntsville in 1985. He was instrumental in developing not only the CAO into an internationally recognized research center, but in promulgating optics throughout the College of Science and College of Engineering, and in the myriad of technology companies and Government organizations in the Tennessee Valley. In 1992, Caulfield moved to the Alabama A&M University and served as its first Eminent Scholar and a Professor of Physics. Later he served as a Chief Scientist at the Alabama A&M University Research Institute and as the Distinguished Research Professor at Fisk University.

John was awarded many international awards, including SPIE's Gold Medal and Dennis Gabor Award. He published numerous books, book chapters, and papers in refereed journals that covered a wide variety of topics. John was featured in several general public publications, including Byte Magazine, Business Week, and the Wall Street Journal which helped to expose optical technology to the masses. Perhaps his prize publication was the 1984 cover article discussing holography for the National Geographic that included a hologram on the cover.

Dr. H. John Caulfield in his Alabama A&M University Office 8 February 2008

John and I first met while working at Texas Instruments in the late 1960s just before he left to move to the Boston area. During the years we developed a fine professional relationship primarily though our activities at SPIE. When the search committee at The University of Alabama in Huntsville contacted each of us as a candidate for the Directorship of the Center for Applied Optics, we each suggested the other while not knowing we both were being considered. When John and I discovered this situation, we said "Why don't we both go and have some real fun together." The committee agreed and the rest is history. John and I traveled together to many places around the world to attend conferences and for work. Traveling with John was often quite interesting due to his occasional directional confusion in going from place to place, particularly at airports and hotels. Now where's John I often asked myself!

From 1985 onward, John and I worked on numerous research projects together that covered all manner of technology fields including signal and pattern recognition, artificial intelligence, optical interconnects, metrology, detectors, fiber optics, object tracking, and even digital and optical cryptology. Interestingly, John and I often consulted for the same companies on the same projects and served on the same Boards. Working with John was always intellectually stimulating and enjoyable. His wide-ranging knowledge base and free thinking often resulted in creative solutions although sometimes the bounds of physics were breached and his concepts were appropriately tweaked in lively discussions. John and I continued to enjoy working together until just before his death. Not only have I lost a wonderful colleague, but truly one of my closest friends.

John lived with his wife and younger daughter at Far Out Farm in Delina, Tennessee. He is deeply missed by his wife Jane, and daughters Cindy Osborne and Kim Caulfield. John will be remembered for his creativity, love of science, noteworthy scientific contributions, willingness to unselfishly help others, and humble friendliness.

Some Personal Reflections
on NES/OSA Past-Councilor H. John Caulfield

Mark Kahan

18 February 2011

(l to r) 1979–1980 President Robert R. Shannon, Technical Director Yale Katz, Executive Director Joe Yaver, Symposia Chair Andrew G. Tescher, Secretary H. John Caulfield, and volunteer Bruce Steiner gathered at SPIE headquarters in Bellingham, WA, in March 1979.

John (Top-Left). Steve Benton (Front, 2nd From Left)

As many of you may know, John passed away on January 31st at age 75^+ from a brief but intense battle with pancreatic cancer. Among John's many many activities and adventures he founded the Center for Applied Optics at UAH, and was Chief Scientist of Alabama A&M University Research Institute and Distinguished Research Professor at Fisk University. He won many international awards, including, among others, SPIE's Gold Medal and the Dennis Gabor Award, and was also past editor of the SPIE journal *Optical Engineering*. When John felt science publishing needed to move forward faster, he stepped up to help enable the on-line Journal Advances *in Optical Technology*. John published 13 books, 40 book chapters, and 255 papers in refereed journals and was featured in many popular news stand publications, including Byte Magazine, Business Week, and the Wall Street Journal. He wrote the cover article about holography for National Geographic in 1984. John got his undergraduate training in Physics from Rice University and his PhD in Physics from Iowa State University. He is survived by his wife Jane, and daughters Cindy and Kim. Services for John were held on 18 February near his home at Far Out Farm in TN, and a fitting memorial is currently in planning.

John was a creative physicist, known around the World for his ability to solve problems. He worked in many fields, including holography, optical signal processing, optical pattern recognition, computer design, artificial intelligence, color perception, and evolutionary psychology. He holds patents on numerous inventions including, but not limited to local reference beam holography, coherence gated imaging, generalized matched filters, optical linear algebra, fuzzy optical metrology, artificial color, and passive conservative interferometric logic gates.

Here, in New England, and before he left to found the Optics Program at UAH ('85), John was both a Councilor and Program Chairman for the NES/OSA, and he worked as a Technical Director in Night Vision at Raytheon, as a Principal Scientist at Sperry Rand working in Coherent Optics, Holography, and Fingerprint Recognition, at Block Engineering with responsibility for all Laser Programs, and as a Principal Scientist at Aerodyne Research working in the subjects noted as well as Optical Metrology.

He loved to think and to learn. He did science because he loved it, and his smile was infectious. Speaking personally, more than just his smile, I found John's combination of new ideas mixed with enthusiasm to be contagious. He was open minded and though he was clearly inventive, he had not even a tiny bit of 'not-invented-here' syndrome. He really listened. I was privileged to work with John over the years, not only through the NES/OSA, but through SPIE Conferences and one-on-one chats. He not only knew how to find synergy in technical things, he found ways to help enable both large groups and small teams to engage and to find new answers. John was a rare talent, rightfully admired by many, and was an advisor to organizations, to companies, and to individuals. Type in H. John Caulfield in any search engine and you are likely to get over 100,000 "hits" that hint at the extensive contributions he made that live on and continue to make a difference. Yet, despite all of the accolades and the real demands on his time, he would always somehow find a way to make time to help someone solve a problem.

A couple of years ago, I called John to see I could convince him to give a paper at the SPIE Annual Meeting. He thought about it and said, he would, but he might not have time to write it up. Well, that's bending the rules a bit, but I thought attendees would really benefit, so the plan was set in motion.

Conference 7796
Monday-Tuesday 2-3 August 2010
Proceedings of SPIE Vol. 7796

An Optical Believe It or Not: Key Lessons Learned II

Conference Chair: **Mark A. Kahan**, Optical Research Associates

SESSION 5: Top-Level Considerations
Room: TBD...Tues. 9:00 to 9:50 am

9:00 am: **When good enough is best**, H. John Caulfield, Alabama A&M Univ. (United States) [7796-14]

John gave his paper (it tied to new ways to use fuzzy logic), and it was very well received and younger members crowded round him later to ask questions and develop a deeper understanding. Recently, in looking over his contributions[1]

---

[1] You can read more about John Caulfield in this 2005 article from *oeMagazine*. (MK Note.- See: http://spie.org/x85438.xml?WT.mc_id=MNLFEB12E and the same article as a PDF at: http://spie.org/documents/Newsroom/Imported/oemAug05/FarOutScientist.pdf)

I noticed that there were 4 refereed papers that he wrote since then (that's since 2010)! He didn't list the paper above. It's not one of the 255 noted earlier. It wasn't formal/official enough. But it was important, and so he did it. For the love it. With John's passing we have lost not only an exceptional talent, but an exceptional human being. But his legacy will live on, and for that, we are all blessed.

Remembering Prof. Dr. H. John Caulfield:
A Man for All Seasons

Marija Strojnik

Centro de Investigaciones en Optica
Apdo. Postal 1-948, 3700 Leon, Gto., Mexico
mstrojnik@aol.com

Abstract. An optical scientist remembers the influences of Dr. John Caulfield.

Professor Caulfield, this is what I always called him, told me that he was probably the scientist who published papers, chapters, and conference proceedings with the largest number of different authors in the field of optics. I suspect that he never really checked this claim, having cautiously included the word *probably*. I am quite sure, however, that his facility to interact and collaborate with many diverse scientists and engineers is the one endearing personal characteristics that will stay with us even if we forget his significant and prolific contributions to optics and related fields of science. We will not fail to remember that he was an incredibly creative and insightful person who would look for a short time at something ordinary and would come away with new knowledge and a novel perspective about it. He was too busy making astounding observations and just asking elucidating questions to have time to put down all his profound explanations. He left us much too early! There were still many more stones to turn and remarkable conclusions to be drawn.

I knew of John from his reputation long before I met him in person. He was the Founding Editor of the flagship journal of the International Society for Optical Engineering (SPIE), *Optical Engineering,* while I was still pursuing my studies at the University of Arizona's College of Optical Sciences. And, yes, I am on that other short list of optics professionals who never published with him. Maybe this is the reason why I feel compelled to provide my own insights about John some that might have escaped even his close collaborators.

I finally met Prof. Caulfield during my JPL-Caltech-NASA years. I have to write it like this, because the Jet Propulsion Laboratory is an overgrown laboratory of the California Institute of Technology where long ago rocket experiments were performed in the nearby mountains. The NASA part comes in because the National Aeronautics and Space Administration has a contract with the University that certain amount of funding would be awarded every year. Thus, the JPL employees were sometimes guided by the NASA rules and other times by those of the University. We felt we were burdened by double restrictions, while our critics felt that we escaped the tough regulations of either institution.

At the time, I was working on a project that was actually funded by a defense agency. I was a senior optical scientist in a group that was involved primarily with studies. My previous experience was as a Staff scientist at Honeywell and a Manager of Optics technology section at a company that today is called Boeing. At both companies, I was responsible for a considerable budget. Likewise, I was quite fortunate at JPL: a defense agency was funding a state-of-the-art project involving optical processing techniques to perform rudimentary vision functions. My task, with a flattering title of principal investigator, was to apply the techniques of the optical information processing to the space robotic exploration, the primary mission of the NASA - JPL. Over twenty years ago, I was performing research to endow a (Martian) rover with intelligence to navigate autonomously, using landmarks that it would see on the horizon. I achieved promising results that an intelligent rover could use optical processing to follow the landmarks under changing illumination conditions. [1] An enormous amount of credit for developing the technology is due to the JPL subcontractor, Perkin-Elmer in Southern California, and the talented engineers who worked there. The current Martian rover, that NASA named Curiosity and ascribed a female gender to it, is equipped with many wonderful cameras, each featuring thousands of millions of pixels. The images, though, have to be sent to the Earth while the rover awaits new instructions provided by the Earth-based engineers. Despite its lack of autonomy, I am actually fond of the 21st – century rover, because the exploring vehicle has gone where neither a rover nor a man has boldly gone before. I somewhat regret that we were not able to empower this mobile laboratory with some intelligence employing in-situ optical information processing or some other novel, self-directing vision capabilities.

One of the wise, conservative policies of the space exploration has always been that technology must have been demonstrated years before it is applied to the robotic space exploration. On a typical mission, such as *Cassini* later on, all the technology development research projects were completed well in advance of the actual start date of the *Mission* and *Project* designations when only the engineering tasks remain to be implemented. [2]

I actually got to know Prof. Caulfield when reporting on such research at technical conferences. For some inexplicable reason, he was always the moderator at the sessions where I presented the results of the research to apply intelligent vision to machines. The most wonderful and memorable characteristic of Prof. Caulfield, for me at least, is his total support of the presenters during his sessions. He always knew what question to ask at the end of my talk, and he never forgot to say something nice when he introduced me. I might have even believed him! I was probably too nervous to respond to him charmingly at that time. John's ease at chairing the session, including introduction of me and finally, his closing of the questions, has taught me how to chair conferences in the past twenty years. [3,4] Looking back, I would say that being a creative and productive scientist is not so much a trademark of this unique, generous, and patient man, as his ease of dealing in a straightforward fashion with all kinds of people. I wanted to say "strange" people, those who do not necessary belong, talking - of course - only about myself. In the eighties, women in optics already existed, but one would see few at the conferences. One time, I was invited to one of the Gordon

Research Conferences that have been conceived to bring together great minds in a certain field to chart its future. Not thinking of myself as one, but wanting to learn from others, I found it a challenge to smoothly integrate into one of those groups that formed one moment and dissolved a few hours later. One distinguished participant even asked me, "What are you doing here?" I am still trying to spin this question into a welcoming remark. But not Prof. Caulfield!

Professor John Caulfield had the human warmth not to see me as a member of different species of humans, who are to walk inside their own rolling, transparent bubble. With our parallel travel through the optical information-processing universe, I slowly gained his personal respect. He usually greeted me at the SPIE fellows' luncheons. Sometimes, we even sat at the same table during those lunches. We talked. I did not have to push on the walls of the bubble to be seen or heard by Prof. Caulfield. When John and I exchanged thoughts on miscellaneous topics, I felt like a person who was a scientist by choice and preparation.

When my own cancer came a few years ago, as a stage IIIc illness, I was surprised because I had always been healthy and I followed all the monitoring guidelines. I shared my warnings with friends near and far. I already recovered by the time Prof. Caulfield experienced his health troubles. After he sent us his cheerful note that excellent medical team was attending to his health needs, I was relieved and hopeful. After a while, I sent him an e-mail note commenting on how pleasant the Sun is in Mexico during the winter months, and how nice it is to take a short walk in the park in the late afternoon. I was hoping that he was doing it too, and that soon he would be telling me about it.

But this was not meant to be.

The field of optics will miss John because his one small conversation can open up a whole new area of research for those of us who put concepts of brilliant people to practice. The humanity will miss him, because his creative mind had many more ideas to formulate and he was always open to new collaborations. I am already missing this exceptional man, because he opened his mind to let me know about the limitless depths of his perceptive thinking. Prof. John Caulfield will be missed by all of us because of his generous respect for anybody who crossed his path.

Quiescat in pace!

Acknowledgments. The funding to prepare this document was provided by CONACYT, the Mexican National Science Foundation. The author's name in this publication adheres to the customs of the Latin culture, including the first name and father's last name.

References

1. Scholl, M.S., Shumate, M.S., Sloan, J.A.: Hybrid digital-optical cross correlator for image and feature classification. In: The 15th Congress of the International Commission for Optics, Optics in Complex Systems, Garmisch-Partenkirchen, FRG. Proc. SPIE, vol. 1319 (1990)

2. Scholl, M.S.: Autonomous star field identification for solar system exploration. In: Mazzoldi, P. (ed.) International Conference, From Galileo's "Occhialino" to Optoelectronics, pp. 802–807. World Scientific, New Jersey (1993)
3. Andresen, B.F., Scholl, M.S., Spiro, I.J. (eds.): Infrared Technology XVII. SPIE Proc., vol. 1540 (1991)
4. Strojnik, M., Paez, G. (eds.): Infrared Remote Sensing and Instrumentation XX. Proc. SPIE, vol. 8511 (2012)

Obituary
Dr. H. John Caulfield

Jumpei Tsujiuchi

Professor Emeritus, Tokyo Institute of Technology

A very sad news came in last February: Dr. H. John Caulfield passed away. I would like to pray the repose of his soul.

A few months before that message, I received an e-mail message from him saying that he entered a hospital for investigating his disease including a biopsy, and I was hoping that he would be able to leave the hospital after the operation even if the disease was a cancer. Unfortunately, however, he had a cancer of the pancreas, and we had finally a sad end.

Dr. Caulfield was always very friendly and kind. Many years ago, I happened to board a plane from Budapest to Zurich together with him, and our seats were side by side. We had a pleasant conversation during the flight, and he went to US and I went to Paris. In this opportunity he gave me a National Geographic Magazine, March 1984, with an embossed hologram on the cover and his interesting article "The Wonder of Holography" inside.

In these several years, he planed to edit some books in holography, speckles, and optical information processing, and asked me to contribute some articles. So, every time I send him drafts of articles, and they were already published. I always admire his passion of developing holography and the related subjects.

Last year, he showed a kindness to nominate me as a candidate of the Emmet Leith Medal of OSA, and I sent him a list of my papers published in English and French together with some comments of the respective papers. Then, he sent to me an e-mail saying "From my viewpoint, you are the Face of Japanese optics. I am honored to call you friend." Unfortunately, however, I was not be able to win the medal, and I was very grateful for his kind considerations.

We have lost a very important parson Dr. H. John Caulfield, and I am very regret his future power and influence in the field of holography and the related subjects.

Author Index